FIELD GUIDE TO SYSTEMS ~~ENGINEERING~~ ASSURANCE

How to Know Your Railway Is Safe

Field guide to systems assurance. How to know your railway is safe.

ISBN Paperback: 978-1-7380016-2-0

To my wife, Carmen, and my daughters, Sofia and Rocio.

You are the light that has guided me through difficult times but also made me laugh more than I could imagine possible.

I love you three.

Contents

Acknowledgments

Dedicated to all those construction passion people I have met during my career and all those difficult but also fun moments we all have shared together. John, Michael, Joaquin, Bea, Andres, Pepe, Ana, Miguel, Gonzalo, Frank, Lou, Jeff, Paolo, Francesco, Giuseppe, Gustavo, Matthieu, Jacobo, Mariola, Jaime, Tomas, Raymond, Manuel, Vicente … I could be throwing names for two hundred pages, and it will never be enough. The world will not be the same without all of you, bunch of crazy fighters, schedule tyrants, and money burners.

I would not change one moment of my life with all of you for anything in the world.

Keep pushing, guys. You are, by far, the best any industry has to offer, all of you.

If anybody who is reading this book has that itchy feeling I had years ago, and desires to move from infrastructure to rail and transit, please find me on LinkedIn. I will do anything in my power to help you, guide you, and even teach you to understand this amazing area that I felt in love with so many years ago.

Ignacio Sanz

Introduction. A View from a Design Builder

This book attempts to explain the complex world of systems engineering to people like me, who finished a degree in civil engineering and went straight into construction management, and to all those others with or without a college degree who have an interest in rail and transit.

This amazing world is full of people such as builders, carpenters, concreters, reo fixers, leading hands, foremen, superintendents, drivers, suppliers, contractors, managers, and many, many others who work very hard to help improve our society's needs but have no time to understand (and have had nobody to explain to them) the complex process of creating and putting a Railway System in service.

I assume you have noticed that I have written Railway System with capital R and S. Strange right? Not to worry. I will explain the reason for it (and it is a personal choice of mine) in a few paragraphs.

Now, straight into the topic with a more common scenario than you think. Picture this: 7:00 a.m. in a project office for a railway project. The construction manager and superintendent are talking with a few team members whose roles have funny names such as assurance director, integration manager, V&V manager, safety (but not construction) manager.

The new team starts questioning our seasoned construction guys about strange things such as requirements (that one sounds easy, right?), verification and validation, configuration instead of change (but I know that change means money and that's it), systems integration and how is it done, because without it, the railway will not operate properly, and many other elements that make no sense at all for our two construction professionals.

And of course, the construction manager and superintendent pretty much answer that they are too busy, and they have a 1,500 cubic meter concrete pour starting in thirty minutes, and sorry, but they need to check safety elements, quality, access paths, final formwork inspection, and many other things, and that the team should schedule another "not so busy" time to talk. And of course, when they leave for the site, they laugh about it and just say that their job is to pour concrete, keep costs low, and ensure that the schedule is met, always with safety. All the "fancy words" stuff is for office people.

Conclusion: frustration on the new team, no interest at all on the construction team, and 100 percent certainty that the Railway project will have delays, overruns, and plenty of rework.

With that (sadly very familiar) approach, there will be situations such as cables required for systems like fire alarms, low-voltage lighting, and many others, that will not fit or will not have redundancy to cover in case of failure. No provisions in the design for maintenance needs when the Railway is in service, how to evacuate people in case of accidents, needs from the local council the railway is going through, and many, many other instances that our two friends don't care about.

That "funny" episode happened to me for the first time in 2014 while I was construction director in the Sydney Metro elevated portion of the System. After all the extra work I had to do after the design was supposed to be finished, with construction well aligned and ongoing according with design, I took a tremendous interest in this lovely world and in fact realize that it is extremely simple to understand and implement.

But as usual, engineers take something very simple and make it extremely complicated, first of all by adding the specific and technical wording to it that, unless you are in this world, you have no idea what anything means. And later, taking the attitude that there is only one way of doing things when, in fact, the same result can be achieved in multiple ways.

As with nearly everything in life, the biggest challenge is communication and understanding each other's needs. When we try to make everything too rigid and exclusive, we end up with these kinds of messes, while with a bit of everybody's help, it is very easy to implement the assurance process satisfactory.

Because at the end, we all want the same thing: putting in service a safe and compliant (contract and regulations) Railway that people can enjoy.

Why Assurance and Not Systems Engineering or Systems Assurance?

Let's start with the basics: what is systems engineering or systems assurance?

For starters, it has absolutely nothing to do with engineering. There is no design on this, no brainstorming on how to design, build, or connect cables, pipes, nothing like that.

So then, why do we use the word *engineering*? To be honest, I have no idea. I guess it is because this whole "process of processes" is done mainly by engineers, and we love to make things complicated.

It can be so detailed, need so much time and effort to link everything together, that it helps immensely when the people directly involved in it have good knowledge of the technical field. In that way, they can add lots of value by helping the designers and builders to make everything successful. At the end, we are all a team, and we need each other to succeed.

Now the second part, why systems? Again, to use a word general enough that it will confuse everybody. It has nothing to do with what the majority of people have in their heads when they hear the word *system*. I bet your mind is going immediately to computers, TV, phones, and in general, anything with cables, chips, software, and similar. Am I right? Well, the answer is yes and no. Those things are normally subsystems of the overall System.

Confusing, right? Don't worry, it is just the terminology. Let me explain.

When you think about it, a system is nothing more than a series of events that happen due to a first activity that starts the whole thing in motion. Making toast could be considered a system: you (trigger) put the bread in the toaster. After a few minutes, you get the toast, add butter and jam on it, and finally eat it. In this cycle, we could consider the toaster toasting the bread, the knife to spread butter and jam as subsystems of the whole system. It is a very simplistic approach, but everything could be seen from that point of view.

In the same way, a railway becomes a Railway System and each one of its components, subsystems. (Never the infrastructure, in my opinion, but we will talk about that later.) All of those subsystems have multiple relations and interactions among them, and all must work perfectly together in an integrated way for the train to run safely.

Here precisely lies the problem for us "concrete heads": systems engineering experts normally have an electrical, electronics, signaling, or software background, and for them, everything is seen with that approach. But for the majority of us, construction builders, who come from big concrete pours, TBMs, launching gantries, burning diesel on excavators and trucks, this is ridiculous.

The Railway world is the only systems-related one in which the percentage between infrastructure (bridges, tunnels, earthworks, drainage, and others) is as much or even more than subsystems (low voltage, medium voltage, traction power, catenary, CCTV, fire alarms, access controls, train control, signaling and many others).

So, what happens? That traditional infrastructure people, concrete-, cost-, and schedule-driven, have to coexist with systems experts with a completely different mentality about timing, when things are needed (much later than infrastructure) and knowing that changes are not nearly always obliging to go back to zero, while in infrastructure it is almost certain that this will be the case. And of course, they never understand each other or make any effort to do so.

And of course, this happens:

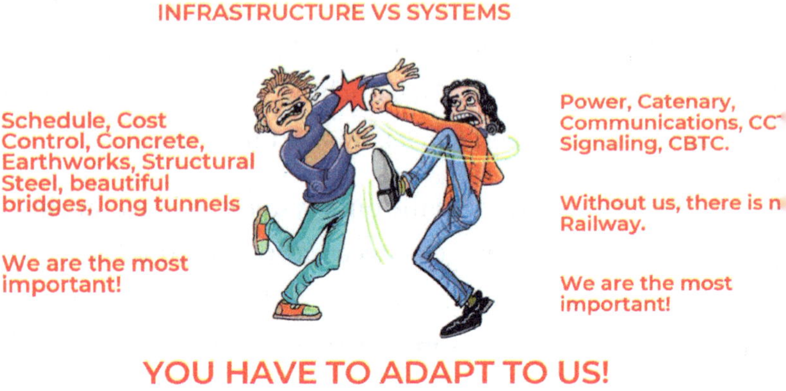

This is the real purpose of this book and my experience poured onto it: finding a middle ground, a sweet point in which both worlds can happily live. Both need each other to successfully deliver the Railway System and avoid those horror stories common worldwide (and more that will be as transit systems are widely designed and built everywhere): years of delay, extra costs on the range of billions, railways in service with hundreds of problems, and an overall population very unhappy because their taxes are burnt on unreliable railways that they still have to use for their daily commute.

To make language available and understandable for everybody and distinguish real engineering of systems (in reality subsystems), I have decided to eliminate the term "systems engineering" or "systems assurance" from this book (and in my career) and call it what it should be called: *assurance.*

I know this new word doesn't help you to understand now, but it will in the following chapters. Just remember, assurance by definition is "a positive declaration intended to give confidence," and this is exactly what this whole process is: to ensure and give confidence that our Railway is safe to run.

To finalize this chapter, my previous promise of explaining why I write Railway System with capital R and S: The Railway is the SYSTEM (this is the only time I will write with capital letters the word System, I promise) we are talking about. One and only one. The rest are subsystems that make the overall and important one work.

Why is the one and only Railway System important? Because of the users, the passengers.

When you drive your car, a bridge or a tunnel allows you to reduce your travel time. When you take one of those, you might think how beautiful it is or how much time you saved.

But when you jump into a train, you don't care about the tunnels or bridges that the train goes though. You only care about reaching your destination because you are going home after a long day, or to a sporting event, your children's function, or something else important for you. You only care about having a safe, reliable, and on-time Railway System.

This is why the Railway System is and will always be with capital R and S: the one and only System which is thought, designed, executed, and operated for users to arrive on time and safely where they want to go to continue with their lives.

Assurance on Railway Systems

As just explained, there is plenty of infrastructure on a Railway, which makes this System unique. Think about another huge and complex system: an airplane. Not much "not cable and pipes" elements on it, right? All avionics, radar, engines, controls, even air conditioning are subsystems, while the only thing that is not a system is what we see, the fuselage that covers it all. Airplanes, which you can compare with the trains themselves, fly through the air (on dedicated paths, of course), airport to airport, but on railways, trains go from station to station using the tracks. And that is the big difference.

In our Railway world, we need different infrastructure (bridges, tunnels, drainage) elements and subsystems. For many of those, we have plenty of already-written standards, local, international, codes, regulations, norms, white papers, and many more, but more focused on infrastructure than anything else.

To make it more interesting, we have stations, which in many jurisdictions have their own code, but in others, there is no specific Railway-related one for these stations. In those cases, we must rely on the local building codes, which are not written for transit purposes. This distinction is key because station users don't behave the same as office and apartment building users (although being the same people in many cases). In stations, people quickly go in and out without paying attention to anything, without practicing fire drills or even without knowing where the emergency exits, extinguishers, or anything are. Those same people know perfectly well what to do and how to behave for all those topics for the office or apartment in which they work and live. Believe me, it is a real headache to make everything work with those conditions and constraints.

What all this means is that we are constantly handling thousands of requirements that need to be complied with, checked, demonstrated, and in many cases, tested. Depending on who you ask, some requirements are more important than others. (Environmental people don't care too much about concrete strength, structural engineers don't bother about signaling, and so on.) But let's not kid ourselves; they are all equally important because failure in one could imply a knock-on effect, which would jeopardize our Railway System from a safety, environmental, economic, public, or even political point of view.

This is the reason why assurance for Railway systems is so important: too many different rules, too many different industries, and most importantly, too many different people with diverse backgrounds and experiences (remember the half-joke at the beginning of this book on the site situation?) with set minds on how to do things as "they have always done it that way." To no one's surprise, each person thinks they are right and use their own language, which the rest don't understand.

With the assurance process, we can create a series of subprocesses that, if properly understood, will help all of us in this Railway world to do our jobs effectively, understand each other, and ensure that we have a safe and compliant Railway system that will achieve what we all want in the end: happy users and—even though nobody wants to recognize it publicly—a profitable business for all parties.

Project Lifecycle

I guess we understand that one, right? Well, depending on your experience, maybe you do or maybe you don't. For those of you who are coming from a pure infrastructure background, it starts with bid, continues with award, execution, and finishes with handover. But of course, on a Railway, this is not the case. Bid and award, pretty much the same, but execution is mixed with testing and commissioning, RAM demonstration, FRACAS (what is this?) and even operations and maintenance.

Why do we need to worry about all those things when we are here to build our project and go to the next? Because of the same reason all over again: to ensure that we comply with all requirements and our Railway System is safe.

Let me explain. A notable number of those requirements (among thousands) that we have are about reliability, availability, and maintenance (RAM). In simple terms, what this means is how the Railway System will react if anything fails (cables, alarms, automatic systems, even drainage blocking that causes a localized flood) and what is needed to recover from that failure quickly or even to continue operations safely while it is fixed. In Railway jargon, we use the term "degraded mode"; when one component fails, what happens to the overall System (System means the Railway as a whole) and how it behaves. All railway systems are designed to continue working in degraded mode to a level of failure; we want our System to continue running when one part of the System is down (depending on what part, of course) but we avoid building failure over failure because it will become too expensive to build if we do that.

In reality, there are so many subsystems working at the same time—hardware, software, combinations—that something fails more often than you think. What will not make sense is that for every single failure, no matter how important it is, we must stop the trains, right? This is where the RAM demonstration comes in. Through statistical analysis, experience, and data from all over the world, we calculate and try to predict how often a failure will happen, how long will it take to repair (including time to detect failure and arrival times), and depending if the subsystem is vital (for example, automatic train control or fire alarms) or nonvital (fare gates or HVAC), we might end up with the need of duplicating cables, pipes, alarms, cameras, or even drainpipes, fences, and others.

So, as you can see, all of a sudden, we are introducing an operational component on execution because without it, we have no idea what we need to design. More cables or pipes means more trenches, excavation distance between elements, and even environmental approvals if we require land not originally available for the System.

Next point: how do we behave in case of an emergency like a fire, somebody having a stroke, heart attack, or something more common such as the train breaking down in between two stations while a big winter storm is happening? How do we ensure that firefighters are able to rescue people? How do we control people wanting to leave the train, and by doing so, avoid them falling from too high (train floor to ballast), getting run over by another passing train, or even losing their sense of direction and wandering along the tracks in the middle of the night? As difficult as it sounds, we need to take all of that into account for every point of the Railway System, and for that, there is no other solution than thinking how the operator will react, as well as firefighters, ambulances, and police. And they all have to be in agreement with our solution and trained properly to react in seconds if something serious happens to our System.

What about maintenance? Normal maintenance, not mechanical or electrical breakdowns. How is the track readjusted, walkways cleaned from debris, snow, ice, or whatever. Fences, doors, stations, access gates, vehicle access to the track, train storage areas, train repair areas, personnel offices, personnel parking lots, and many others. All of this must be considered during execution, and there is no other solution than engaging early with the maintainer.

Your contract will not specify every single requirement that all these stakeholders will have and no other solution than yet again talking about assurance. In this case, requirements management, which we will be covering in one of the future chapters of this book. For the moment, just remember than in a Railway System, execution is not as straightforward as you have just seen in infrastructure projects.

To make it even better, we must ensure that everything works together: testing and commissioning. Not only each independent subsystem on their own but all together making the Railway System.

A standard and known representation of all this explanation is the V diagram:

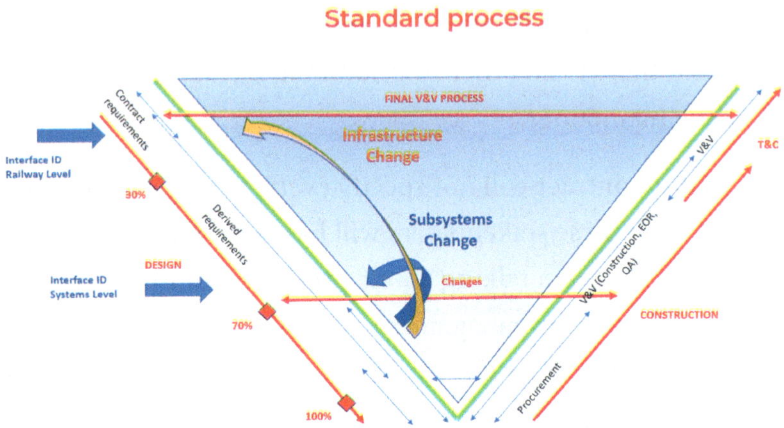

What it represents is that all the requirements—identification, design, interfaces, and discussions—happen on the left side (coming down the V) with construction happening on the right side (going up the V). T&C and the end of the cycle.

The two horizontal arrows represent some changes that could happen all along the cycle, even during the final stages.

Yellow and blue arrows on the left are my personal addition to represent that changes on infrastructure will normally affect the design more than changes on the subsystems, in terms of design time and having to go back to early stages.

Looking at it seems simple to understand, but the problem is that this diagram is just theoretical and based only on the existing standards for the process.

The reality, sadly, looks like this:

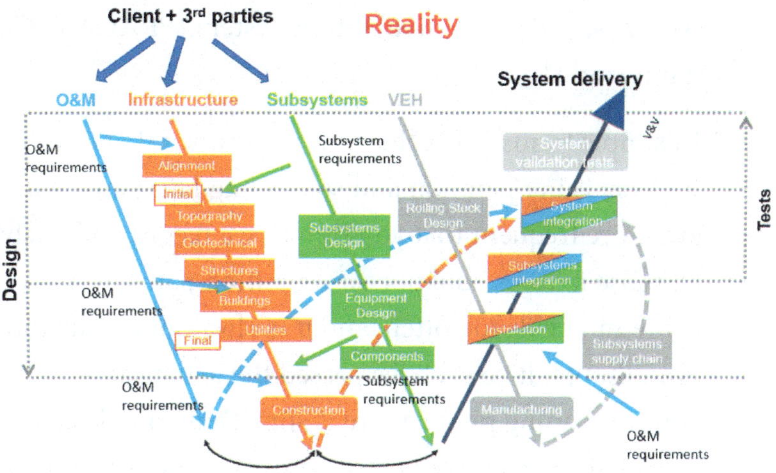

What this confusing diagram is representing is nothing more than V lifecycles for the four main components of a Railway System (O&M, vehicles, infrastructure, and subsystems), all of them intertwined, with requirements coming from many external stakeholders, and in between them all the time (even when we go up the V, during construction). This is the reality because each party has different timelines and needs.

Without being extremely organized, we will end up in a complete mess with delays, penalties, and the usual problems we must avoid.

But this is not all, because finally, we also have (not represented on the diagrams) the Failure Reporting, Analysis, and Corrective Action System (FRACAS). Cool name, right?

In simple terms, what this jargon means is that once the Railway is in operation and something fails (small failures are more frequent that you think), we record and analyze all those failures, looking for repetition patterns. If something fails too often, and it is always the same thing, even though it could have passed the testing and commissioning phase, there is something not designed or executed (or even connected) properly and is making the System fail.

With this FRACAS process, we identify that and establish corrective measures, which not always mean a repair but maybe a component or even subsystem additions to correct the repetitive failure.

And this is our responsibility as design – builder for those elements that we have built. Our job doesn't finish after handover, but continues during operations for a period of time, normally till the end of our subsystem's warranties.

To make all of this even more interesting, nowadays our clients tend to award Railway systems on the order of billions, with many lines (meaning here train service lines) at the same time, mainly to ensure having the same vital and nonvital systems consistently working everywhere. And due to the increasing complexity of everything we do, seasoned with public opinion and political conditions (always present on transit systems), we are requested to also open the System progressively (one line each time) rather than everything in one shot.

What this means is that from the moment we are awarded the contract, we have design, execution, testing and commissioning, handover, operations, maintenance, FRACAS, and even the possibility of emergency situations, all at the same time, before we are 100 percent completed on everything that our design-and-build contract mandates us.

Like the 2023 Oscar-winning best picture title says: everything, everywhere, all at once.

Organizing the System. SBS = System Breakdown Structure

Another new term. What does this mean? System Breakdown Structure?

Let's go first with what we all know: work breakdown structure (WBS). I hope all readers of this book are familiar with this. But just in case, a WBS is the traditional way of dividing an infrastructure project in areas that can be tracked independently in terms of schedule, cost, resources, safety, quality, etc.

What this ends up meaning in practical terms is that this division is geographical and has a project manager in charge of every aspect of each area. Simple and easy to understand, right? The project manager is responsible for every single aspect of a job from this point in the map to that point in the map. Clear boundaries, responsibilities, reporting, safety, procurements, quality, construction, and everything else. Job done?

Well, not quite.

Imagine this: WBS of Station A, different from WBS of Station B, two project managers, two teams. And each one of them has the same systems (fire alarms, HVAC, ventilation, lights) but different procurement, so different subcontractors, and later, all subsystems must talk to each other. It will never happen, no matter how hard we try, because different suppliers are not prepared to interact which each other by default. On top of that, when you try to find all the information required for a particular subsystem such as all fire alarms for the overall testing, it is impossible to find, as the info is organized by geographical area. Each station can have their own fire alarm system working perfectly on its own, but when they must work together, it doesn't happen. On the overall System, when one fire alarm kicks in because there is a fire at that station, it must indicate to the others that there is a fire and to the trains as well, so they don't stop at the station on fire, instead stopping before reaching it.

Imagine yourself in an automatic training going nonstop to a building engulfed in flames. Pretty, right?

Another example: you must identify interfaces between subsystems for each station every single time. For two stations, it's not bad. What about for twenty-four stations or more? Same logic for portions of alignment between stations, which is another typical WBS way of dividing the job. Are we going to do the same exercise multiple times?

It is a mess, because all linear requirements are the same. So, are we supposed to repeat the same exercise as many times as the areas in between all the stations we have? It makes no sensc.

The madness, the horror!

Here comes the SBS to the rescue. There needs to be a code for each subsystem, so it is easy to identify all the documents related to that particular subsystem or to create interfaces between subsystems. And at the same time, keep the WBS too, so all traditional cost control, scheduling, safety, and everything else remains familiar, and everybody can continue doing their job.

This, in practical terms, translates into creating a document control coding that includes both WBS and SBS, and it is enforced in the whole project, consultants, suppliers, contractors, and even the client.

We will come back to why SBS is vital on a Railway System when we talk about testing and commissioning, as well as the verification and validation processes.

One final but very important note on this: SBS must *always* be created during the proposal. I have seen too many projects go by WBS, and the effort to renumber everything later makes the whole project run six months late, creating confusion, frustration, and upsetting many people about it when there is no need.

Roles and Responsibilities

“Not in my position description.” Sounds familiar? I bet it does. These days, this sentence and similar ones are becoming more the norm. Behind us are the days of people doing everything on their own and solving all problems. Apart from not being healthy for anybody, it is only prone to errors. We do not need superheroes, we need teams. People working together for the same goal, with clear roles and responsibilities and motivated for the success of the group and not the individual.

How do we do that? The first approach is to ensure that there are no scope gaps on what is needed to be done. When recruiting, the position descriptions must be specific and detailed of what is required from each individual, but we also must make it clear that it is not a “must” to be able to do everything perfectly. If we do that, we will never find a team. In this assurance world, we are very few, and we all must be willing to constantly learn but also constantly teach. We are the ones to do the interviews and find the potential, the willingness to learn, and the ability to understand.

Hopefully, this book will help make this language we use clearer for everybody and will increase the number of us in this amazing field.

I am going to try to explain now, with easy-to-understand language, the roles and responsibilities of the main players in these processes, the way I see it could work. Brace yourselves, because it will not be a very fun ride, but I hope it will be an eye-opener for you.

- Documental Control Manager: responsible for documentation to be organized in accordance with SBS and WBS and that all workflows are properly implemented and applied by all parties.

- Environmental Manager: responsible for early engagement and ensuring that environmental teams defined SMART (Specific, Measurable, Attainable, Realistic, Traceable) requirements, which must be considered as rigid as any other contract requirement and applied the whole assurance process to them.

- Quality Manager: on top of what we all know and understand about what a quality manager does, it is very important to ensure construction proof (ITP, inspections, audits), and in general, any quality-related paperwork is extremely well recorded because for infrastructure elements (concrete, earthworks, drainage, etc.), these will be our validation proofs to ensure a requirement is met. They must be completely aligned with the document control manager and be the biggest pain for our construction colleagues. On top of that, tracking of NCR and deficiencies is key, because any one of those being left open could impose a safety risk that will have to be studied to decide (and demonstrate) if it prevents the Railway System from operating safely.

- Configuration Manager: one of my personal favorites. The key role to understand the effect of changes on the whole design. To be explained in far more detail in a later chapter.

- V&V Manager: the champion for verification and validation. The police to ensure we all provide everything that is needed to close the process.

- Requirements Manager: database custodian. One of the first ones to be engaged and in charge of ensuring that everybody knows what to do at all times and which requirements apply to who.

- Assurance Director: little "know it all." The one who knows how to do all the previous roles and ensure that the orchestra plays the music coordinated. Creator of all procedures and ultimately responsible for interpreting the standards in a practical way for the whole team (assurance and outside of it too) to apply. Professor for the "new pupils" but also for the uninterested dissidents from the process.

- Security Manager: both physical and cybersecurity. The one to identify possible security threats to our Railway, propose mitigations, and audit that those mitigations are implemented for closing the threat.

- Safety Director: identifies possible hazards affecting safety of users and proposes mitigations. Statistical analysis based on experience and putting together the final safety case.

- Testing and Commissioning Manager: sounds interesting, but it is a nightmare role. All testing of all kinds and its proper documentation are their responsibility, as well as ensuring that all subsystems work together.

- Interface Manager: identifies interfaces, both internal and external, and creates requirements from those.

- Design Manager: ensures that design complies with *all* requirements and takes responsibility for it. Fun, isn't it?

- Engineering Manager: this one is even more fun. On top of helping the design manager, the engineering manager also needs to ensure that it is "buildable," on budget and on time. Also, full integration with temporary works design (formwork, scaffold) that could end up changing permanent design and even affecting requirements.

- Technical Director: the craziest of all—and at the time of writing this book, me. Responsible for all the previous ones with the usual commercial, contractual, and client relationships. Usually the "bad cop" within the organization, as it is the one telling people "No," many, many times because it will go against the process. Usually, the most thankless job and very demanding but extremely rewarding, as you see everything forming up according to plan. Who doesn't love when a plan goes well?

As you can see, I have not gone into lots of detail for any of these roles, as each one could have its own chapter. I just wanted to highlight that without all the players, the task is a huge one for a small team, and the project will fail. We cannot underestimate the size of the assurance group, but also, we are smart enough not to create a small army (which many systems assurance experts will do and say it is "the only way").

Requirements Definition. When and Why?

We finally cut to the chase. First stop: requirements definition. One could think that simply reading the contract and applicable standards, we have everything we need, but sadly, it is not the case. Contracts are written, on many occasions, with no clear language, and requirements cannot be general but specific. That language is good for lawyers but not for engineers. As said previously, requirements must be SMART, which basically means that they need to be measurable and concise. We cannot have things like "be safe" because it is subjective, or "reach agreement with stakeholders" for the same reason.

Of course, contracts are not just a list of requirements perfectly organized in a way that everybody can understand them. That will require the use of common sense, and as we know, that is not what we do in this business, do we?

Out of the five thousand pages (probably much more) that our contract might have, we are going to be the ones needing to extract requirements from it. So, do we get somebody to read it and manually write, classify, and make them SMART? Believe it or not, a few years ago, that was the way(note to Mandalorian fans). These days, there are tools and software that can do a first pass for us. These database tools (in many cases with a level of artificial intelligence built in, to make it even better), created specifically for this task, can read the contract, and based on some key words, break it up into requirements. That first pass always has to be reviewed by the requirements team, but it greatly reduces time.

Later, those requirements must be classified based on fields, that these days are normally known by the requirements experts but always must have SBS (now it starts to make more sense), design discipline, type of requirement, owner/leader, status, and priority. It is very important to also have the "artifact" or the way of demonstrating the requirement, but we will go into more detail in the verification and validation chapter.

As painful as it looks, this is only the beginning of the story. These contract requirements break down, and classification must be done during the proposal so we can start with a clean database for the teams to work. Ideally, the elimination of general requirements and translation into SMART ones should happen during that phase too.

Two personal opinions here:

1) Never decompose standards. Some requirement managers tend to believe that to make everything even easier, it is better to decompose standards so

everything becomes lists with bullet points, easy to track. While this could be true in nearly the whole world (although not recommended because it creates unnecessary work, as engineers already know their standards), it is the wrong approach in North America, due to liability obligations of the professional engineers. This will be discussed in one of the final chapters of this book, but just keep it in mind for the moment.

2) Disregard nontechnical requirements. Lovely discussion point on every project. This assurance process must apply to technical requirements only, but for many "extreme" requirements, people want to use their magical requirements database for everything. I am talking about requirements of the likes of submit monthly report, answer social media in twenty-four hours, finance submissions, safety statistics. Although it would be nice to have everything in one place, the trick is that you are making every requirement to be traced to the same level of scrutiny and in the closing stage, having too many of those becomes a burden that nobody wants. Follow my advice, and have the fight at the beginning, to pass all those nontechnical requirements to the relevant responsible department and let them use other

simpler software tools such as Excel or Access to track those.

And we continue. Once we have the contract requirements (and never the standards decomposed), what comes next?

Well, as crazy as it sounds, we simply need to find more requirements. At the beginning of this book, we explained that Railway systems must be compliant and safe. This last word is the catch: *safe*. And here is when all hell breaks loose: interfaces between disciplines, subsystems, and software to ensure that they don't create conflicts in our computers between them, protection in case of emergencies, fires, passengers with reduced mobility, visual impaired, hearing impaired, access for firefighters, ambulances, police, conditions from local councils, provincial authorities, external regulators, environmental, community concerns, and even suppliers who cannot do exactly what we ask of them but can do something similar. And to make it even more fun, what the company running the trains later will require to maintain trains, infrastructure, assets and how to operate all of those.

Ideally, all of those will give us SMART requirements very early because everybody is crystal clear about what they want and how to do things. (Did you notice the sarcasm there?) The reality is that they don't. Local authorities have limited resources that are stretched too far, firefighters explain what they need but don't indicate how we must do it, same for ambulances and police. And operators and maintainers are never with us early in the project, for the simple reason that, same as us, they need to protect their costs, so the number of available resources and their knowledge is very limited.

None of those people's jobs are assurance, so the same as our infrastructure specialists, they are not thinking about requirements, assurance, or this craziness. They have their own priorities, and as frustrating as it can be, it is understandable. It is up to us to explain, ask, insist, and in the majority of cases, hold their hand while asking specifically what we need and why.

Quick example: If you build a maintenance road parallel to the tracks logistical preparations can happen safely during the day, while the Railway is operating, and later at night, the maintenance or repairs can be done, but that road can be used for construction access, evacuation, firefighters, ambulances. But on the other side, the road costs extra money to build that maybe the client doesn't have. And depending on the type of road (dirt track vs. asphalt) its maintenance could also be expensive. This simple example happens on every project and decides too many things around it. If you must focus on a particular request, focus on this one. And drive everybody on the conversation, indicating that if the road is not there, maintenance will lose time preparing and arriving at the required spot, and emergency situations will have to be covered by the operator or with another infrastructure that probably has not been thought about. Insist on it until all players give an answer.

As usual, we want everything as early as possible, to avoid having to come back on design or reconstruct, which is sadly far more normal than it should be, due to not knowing what is needed. Whatever we don't get, we will have to make assumptions that we will treat as requirements in our database, so we have something to go on. It is very important to communicate to everybody what our assumptions/requirements are, to avoid complaints, surprises, and claims from anybody at a later stage when the experienced personnel come into play.

Who is going to tell you what you don't have? Tricky question. It will have to be your team and you. That is why having people with the right experience and asking many questions is the most important thing in this whole process. Those roles we described before are key to be filled at the very beginning, mainly for this assumptions/requirements process, when you cannot (and will not) obtain all the answers when you need them.

Remember, every effort at the beginning of the project will make everything smoother and reduce frustration for everybody in the long run, even at the cost of your personal frustration for having to explain everything to everybody multiple times to get something close to what you need.

Believe me, it has happened to me far too many times and still does. But it is the game, and a very interesting one. And remember, tell the world what you are doing (in writing) and why. The "design assumptions" document has helped more times than I would like to admit. This document will oblige everybody to think, and chances are, they will come to you with requests for more time, requirement changes, or allowances for the "just in case." All of these situations will play in your favor in the long run.

To help a bit, here is a visual representation of all requirement types as general categories.

Requirements Definition

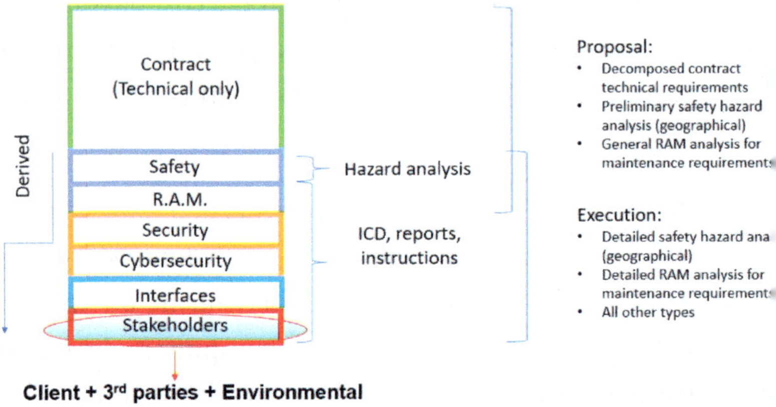

Role of Safety and Security. And RAM~~S~~

First things first. Why do I write RAMS if later I am crossing out the S? Again another personal opinion.

RAMS stands for reliability, availability, maintainability, and safety. And this was correct, as it is how it started years ago in systems assurance. But in my view, safety has grown so big and important that it deserves a treatment on its own. And it also now covers security and cybersecurity. So, I don't use RAMS but RAM, safety being its own discipline, although very likely managed by the safety director anyway, with their own RAM specialists in the group.

Now that we are clear that we are dealing with two very distinctive topics, let's go with the first one: safety and security.

Safety is even more important than compliance. No matter what our contract requires from a technical point of view, if our Railway System is unsafe, nobody will ride it. Common sense.

How do we know it is safe? The combined experience of the team on similar jobs. Once we know the type of Railway, the environment in which it is going to be operating, passengers or goods using it, we can identify the hazards that could affect it. Is it going through an industrial area with lots of gas tanks? Over a wide river? Elevated? Underground? Will it transport dangerous goods? These are only a few examples of situations we can have. But then we must also think about emergency scenarios: how to deal with a fire on the train. How to deal if it happens in a station. How to evacuate people. How emergency services will access if somebody has a heart attack. How to cut power if needed in a quick and safe way. And many, many others. This is why the combined experience of our team is key. Identifying all potential hazards, or at least as many as possible, will allow us to propose mitigations to solve them. Those mitigations could be procedures to be implemented in operations, which is always the least desired solution, but also new things to add to our infrastructure. Some examples: access doors that can be operated remotely, emergency lights, designated areas for people evacuating to walk, secure rooms in stations to wait in case of fire, backup generators, extra smoke-extraction systems.

Many of those will not be in our contract. If they are identified late during design, it could mean having to start from zero. Imagine we discover that in an elevated part, we need to add a designated area for people to walk during evacuation, and that area means our deck needs to be wider. Now imagine that deck was posttensioned, and we are already about to start construction, so design is nearly finished. This could mean that even the piers could have to change, and if they were already in construction … better not to even think about it.

That example could seem extreme, but it has actually happened on a job. And worse in some cases. Don't diminish the extra infrastructure that safety requirements can add to a job.

Just as the recommendation I gave you with the maintenance road: always start with how you will rescue passengers in case of fire in every point of your System. Decide early whether to use infrastructure or operational method and leave it in writing with your client and your operator to avoid legal disputes later on. This is always the main source of conflicts in every single project.

When you have identified the hazards and a few acceptable mitigations for each one, take them and negotiate which one to apply for every single case. And if you have a safety committee, which is the most common circumstance, this is even more important. Having the agreement from the involved parties, and if possible, those extra requirements negotiated with your client early, will prevent multiple headaches later, when money is tight and nobody wants to commit. So, take all those mitigations, present them to the safety committee, explain which one you prefer, and let the committee agree with you or not. If they don't agree, make them choose one and record it. Write, write, and write.

Now you are ready to add all those mitigations to your requirements database. And this is key: Add not only the ones that oblige you to build something extra; also add the ones that have been decided to use an operational procedure. It is not for you to do, but you will need that information later to demonstrate that your Railway System is safe (you have a mitigation for each hazard) but also for commercial disputes that, believe me, will come.

But I have bad news for you. More hazards will come later during the project design life. Sadly, it is nearly impossible to cover it all. We are humans, and we cannot cover it all. But if you have covered the most important ones, you will find that the new hazards will likely be solved with mitigations that you already have in place. Killing two birds with one stone!

If not, there would be no other solution than to adapt what you have to the hazard mitigation identified and pray that it doesn't involve too many changes. That is why an experienced team from the very beginning is paramount.

Security is the same approach as safety, but with topics related with terrorism, trespassers, bomb threats, and the like. And here is where it is difficult to put a limit. Do we need to design for a bomb explosion? That would be extreme. What we must consider is mitigations to reduce that possibility as much as possible. An expert in this field and police collaboration are key. Same as with safety, early involvement is needed to avoid surprises. In many cases, CCTV and fencing are acceptable solutions to cover nearly everything. People can avoid them, yes, but not as easily as you think. Having a security consultant on your team, reporting to your safety director again, early, will give lots of light on the situation. For security, the hazards identification is done the same way as with safety, but with a much more reduced group of people (we are talking about very sensible information that identifies vulnerabilities that criminals could exploit) and it has its own name: threat and vulnerability analysis (TVA). In this case, the TVA will not be made public, only the mitigations, in the form of design additions that we need to consider. Again, straight to the requirements database and subsequently to the relevant discipline for designing it.

RAM. This is not related to safety but to having the System operating as much as possible and with any repair or failure affecting it solved in a short time. What happens if one of our power cables is chewed by a rodent? Does the whole System go off and everybody gets stuck in the middle of wherever they might be? Not acceptable, as you can imagine. The same could happen with fiber-optic cables carrying all our data needed to run the different software, pipes for water, or pretty much anything you can imagine. The simple solution is to have everything duplicated, which is what it is required for all the vital subsystems (the ones that are involved and required in saving people's lives), such as power, fiber optic, fire-extinguishing systems, fire alarms, access control, and others. But you will not duplicate piers on a bridge just in case one fails. For that, you design with required loads, durability, and impose some maintenance on the structure to ensure its required characteristics over time. It is another way to ensure the RAM of the infrastructure. And of course, even the duplicated subsystems will need maintenance, and it is up to us to indicate it. We will also explain that in a later stage.

But we also must consider the time needed to repair anything and how to do it. How and how long will it take to replace a concrete tie? A rail? A tamper ballast? To replace cables? Fix an alarm or a camera? We must consider every single element of our design and calculate, based on experience in other Railway systems, the time it will take to repair or replace them, their location and environment. It is not the same to repair an element that is five kilometers away from an access as one that is twenty meters from it. Just as it is not the same to repair in summer as in winter or night shift versus day shift. We must consider the worst-case scenario and plan for it. Again, this will create more requirements that will have to be added to our requirements database and considered in our design.

So, yet again, the sooner the better, and if the operator and maintainer are not telling us anything, we must use our team experience, record it, submit it to the client and operator, and include it in our requirements database. Just as we have done with hazards and security mitigations.

I also recommend your RAM specialists report to the safety director, as with the security specialists, because many requirements for RAM might already be covered by your safety requirements. You will only have to explain how to use what you already have, to cover the RAM requirements.

Now you see why at the creation of this discipline, it was called RAMS. All together can cover all requirements that are not on your contract or standards, and they can be complementary between themselves. But due to the importance of safety over the others, I have decided (and recommend you do the same) to separate them. Start with safety, especially emergency situations, and go down from there to security and later RAMS.

Life will be so much easier that way, rather than trying to do everything at the same time.

To help you a bit, below is a simple diagram on how to identify safety and security hazards and end-to-end path till reaching a safety case. Not everything makes sense now (V&V and safety case), but I promise you will understand very soon.

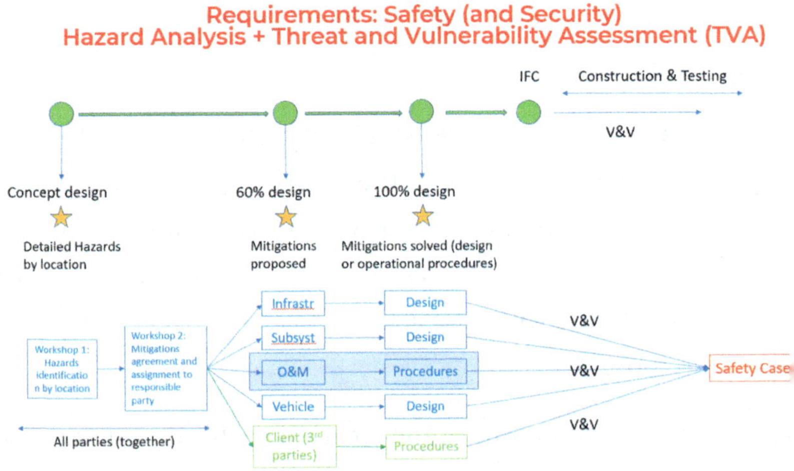

Cybersecurity. Why Is It Different? What to Do with It?

When we talked in the previous chapter about security, I purposely omitted to talk about cybersecurity because it has a very different treatment.

Have you seen the movie The Italian Job? I refer to the remake with Charlize Theron, Edward Norton and Mark Wahlberg, not the original one with Michael Cane. In that 2003 movie, there is a car chase in which the thieves team hacker switches all traffic lights to green, so their companions can cut through traffic, and when the bad guys are chasing them, switch them all back to red. You know what happens then. That is fun to see in a movie, but imagine the situation in real life. And now imagine that somebody goes into our Railway System software and starts tampering with train controls, power cutoffs, access control, emergency lights, or tunnel ventilation systems. Imagine that person increasing train speeds, disabling brakes, all of that remotely. Or something simpler: one of our operators bringing a pen drive with the latest memes, and on those, we have one or many viruses that put the System down. Not fun, right?

But contrary to everything we have discussed until now, for cybersecurity we don't have very prescriptive standards because it is a world that changes so rapidly that by the time we want to create such standard based on a procedure, somebody has already found the way of hacking it, and then we need another security update. Pretty much what constantly happens with our phones, computers, and everything that has any software in it.

Then what can we do? Adapt. Your cybersecurity team will start by creating a number of requirements for all of your vendors to comply with. Because with software, rare is the time in which you design it yourself. You will (nearly) always will go to the market to get each subsystem provider to work on their specific software, and you will be the one to ensure that all that software can work safely together. That is the job of your cybersecurity team/specialists.

Different vendors will be able to comply with some of those requirements, or even better than what you want, and your specialists will have to go through the proposals from those vendors and see if they can accept, reject, or adapt. This last one is the key word: *adapt*. Maybe a particular vendor cannot do what you want but can provide an acceptable alternative that ends up solving the problem you were trying to prevent in the first place. That is the reason behind cybersecurity being different: you don't do it in house; you must involve your team in the whole procurement process, and you will need to adapt.

There will likely be lots of rules you will have to impose on the operating team, such as not to connect any external pen drive or hard drive on the computers running the System, not to leave doors open at any moment (it sounds stupid, but it is a more common problem that it seems), and not to allow browsing the web with those computers or even with the dedicated Railway internet connection. (The safest software system is the one not connected to the web.)

There are discussions about having the cybersecurity requirements in your overall requirements database or not, and in this case, I have not made myself a definitive decision on it. On one side, these requirements will adapt and change so much that having to do a configuration-management process (changes control, which we will talk soon about) every time could take lots of time and in reality not much knock on effect on physical requirements, except in a few isolated cases. But on the other hand, having them in the database gives a full and complete picture of the situation, and although painful to control and adapt, gives the whole picture and more traceability, especially in case of audits.

Call me old-fashioned, (which I am a bit more every day, same as you), but as I am not the one having to write everything down (I know what you are thinking, but I am simply honest about it), I like the second option more, although I understand why some specialists prefer a dedicated and isolated approach until the end of the project.

External Stakeholders

We have done a lot already, but everything until now has been "in house." We have only discussed our contract requirements, what we find that is unsafe, unsecure, hackable, and how to solve it. Also, we have considered breakdowns, repairs, and if it is a good idea to duplicate elements. So far, so good.

But we have not gone to the wild world.

With airplanes, airports are more or less isolated, and planes have flight paths. The external influence is more or less controlled and known, apart from being extremely regulated.

With Railways, it is a completely different situation. Our track goes through municipalities, towns, cities, provinces, states, and even different countries. Through forests, over rivers, bays, underground, elevated, at grade. Possibilities are nearly endless.

Each one of those has another organism with its own rules and regulations, which translate into requirements that we also must consider. Just because we are building a railway that will improve people lives (that is always the goal) doesn't mean those regulators will "close their eyes" an allow us to do anything we want. To make it even more interesting, remember that you are never their priority. They must deal with their normal business, and you are nothing more than another applicant. Bigger but same as any other. This translates to never enough resources, with less knowledge on Railway systems than you have and no time at all to answer you quickly.

Pack a ton of patience. It will be on you to make every single proposal, every single presentation multiple times to different people and repeat it again and again until you obtain an answer.

And mark my words, they all want something in exchange: a bigger stormwater system because there is municipal planning about to happen, addition of traffic lights on an intersection, existing road bridge upgrades and others. Because you are going to need them to connect multiple services, divert traffic (permanently or temporarily), get extra land or any other request, that will be what it is needed to make it happen in accordance with their local regulations. And your client would have not dealt with everything in advance, as it was supposed to be, because lack of resources, knowledge, or any other reason. It is impossible to plan for everything.

I am talking of cities, councils, towns, utility owners, road authorities, building authorities, other railway operators, and emergency services (firefighters, ambulance, police).

You are affecting all of them by adding something to their normal life, so you will have to listen to them and comply with their requirements, or they will not allow you to build or enter into service.

Very important: again, record everything as early as possible. Management of these entities is another typical and repetitive source of disputes with clients, because more often than not, your client will not have an amazing relationship with all of them, and they will use you as another reason for their discrepancies. You don't want to be caught in between two fires! Same solution as always: propose, present, and record in writing. And as early as possible. Never underestimate engaging third parties and creating good relationships from the beginning. Good relationships always pay off in the long run.

Once you have those requirements, add them to your requirements database, but remember, never decompose standards. Even if they are local, your designers will know them by heart (or they should) so there is no need for you to do that.

Interfaces between Players

Players include designers, operators, and maintainers. Ideally also all the third parties we have just described.

Let's talk about designers first. Different specialists care only about what they know, and sadly, unless you tell them what to take into account from the others, they will not do it on their own. For example, structural designers may not consider the bolts for catenary posts unless you oblige them to do so. Same logic for track or drainage, cable containment (number of cables to consider from all parties), power requirements, and any discipline you can imagine with each other.

Now, there are two approaches here: you tell them in detail what to consider or you make them talk to each other. You obviously do not want to be responsible for the first one, so we go for the second: make them talk.

How do we do that? Interface matrix.

An interface matrix is nothing more than a matrix in which you write all infrastructure disciplines and subsystems in your Railway System, and you mark which one has influence with the other. Drainage will not have influence with train control, for example, but it will have with earthworks, track, and utilities. Again, you need the expertise of your team to identify which discipline or subsystem has any influence (interface) with which other.

Once you have identified the interfaces between them, then it is time to assign a leader, who oversees developing in detail all the interfaces between those two. Another piece of advice: choose as leader the one who needs the information first on your design schedule. Example: interface between a catenary system and bridge design. The bridge designer will be the leader, as they go first before catenary, and they need to know catenary post locations to take the bolts that fix those bolts to the deck into account for their design.

This is one of the points of difference between North America and the rest of the world. We will cover this situation at the end of this book. Just remember that liabilities and responsibilities are key on this point.

Remember what we said a few chapters before: *infrastructure is not a subsystem.* Don't create "artificial interfaces" between infrastructure elements such as drainage with earthworks, structures with drainage, and similar. Interfaces must be between subsystems or between a subsystem and traditional infrastructure discipline. Examples: Catenary – Structures, Signaling – Track, Traction Power – Cable Containment. A very important note: many systems engineers will try to convince you at least to consider track as a subsystem. In my opinion, it is not, and the simple reason is that track discipline, just as with the other infrastructure disciplines, has a well-defined set of standards which cover nearly every aspect of them.

Now that we have identified all detailed interfaces, we must record everything (as usual). The document is called the Interface Control Document (ICD), which both parties must sign. It is in everybody's best interest to include all requirements as SMART requirements (remember what we explained a few chapters ago) in the ICD. More advice: do not accept when people write things on the ICD such as "it will be developed at a later stage" or "enough to …". General language should never be accepted.

The reason behind this is to take the ICD, once signed, and then give them to our requirements team to extract requirements from the ICD, include them on the requirements database, and assign them to relevant parties (which will be the ones who signed the ICD in the first place). Later, for the requirements assigned to you, give them to the relevant design lead (per discipline) to be incorporated on the design.

Ideally, you should have an ICD with the operator, maintainer, and all those third parties we talked about.

When I was repeating about write and record, I was referring to ICD. Making everybody sign these documents will make them accountable, and they will end up studying what they sign before they sign it. If they don't sign them, produce—based again on your team experience—an assumptions document, which you will send to the parties to which it applies, indicating which of those requirements will you consider or not, and of course, why.

Remember to always project yourself!

Interface control documents are one of the most difficult parts of this job. Nobody wants to commit to anything because overdesigning or underdesigning will bring hard consequences for all of us.

The problem is always the same here: for subsystems, changes in a later stage will, a majority of times, imply software changes, more testing, and barely anything else. The number of cables as well as their diameters are normally known since the beginning, and on the standards, there are always requirements for extra capacity, to cover these situations. On top of that, clients always add more in their contracts, so there is always the possibility of asking for derogations to cover extra unaccounted needs.

But in infrastructure, we are in trouble. Nearly any change that happens in a later stage means that we don't have enough space, load calculations are not valid, foundations insufficient, slopes and geotechnical limits don't allow for corrections, we didn't consider extra penetration on fixed elements, concrete doesn't allow for more reinforcement, extra strength, or God knows what.

This translates into the infrastructure team wanting to have details as early as possible, but subsystem teams pushing back to avoid over- or under-asking because details don't come until their own design is further advanced. They need lots of computational simulations with plenty of variables, which are not the same depending on ambience, type of infrastructure, use, and even frequency. And of course, infrastructure design always goes first, which makes the urgency even worse.

Here is where the technical director will make the final decision. It is their role to consider higher assumptions than real ones (or not) and take responsibility for whatever comes later. Not an easy part of the job, but needed, because if not, the cost of rework and delays will always be worse.

For this one, I am sorry, but there is no advice I can give you. Experience it, learn from your errors, and the most important of all, surround yourself with a group of trustworthy professionals and listen, listen, listen before making a final decision. But do not take too long, or the schedule will eat you alive.

Finalizing with a bit of help, what does an interface matrix look like? Well, depending who you talk to, they will tell you that the next example (which is my personal general matrix) is far too general, and they prefer to go to the detail about which cable, which connector, which bolt, but this is not what we are doing in this book.

For now, just have a look at the matrix, which is just a template.

NOTE: due to size of this matrix, it is not included in this page. Please refer to Appendix. The graphs on page 129, at the end of the book.

The way it works is as follows: all possible disciplines or elements that can interface with each other are repeated on the top row and the first column on the left. Now there are two approaches: simple or more complex.

- Simplified: On one side or the other of the black diagonal (so one side will not be used) just mark a X in which an interface between two will happen. Then each cross will mean at least one ICD to be developed to define clear SMART requirements for later input into our requirements database and assignment to the relevant designer.

- Complex: Select either row or column with discipline/elements to represent the interface leader. Depending on who you want to lead ICD creating X could be on one side or the other of the black diagonal. What this allows us is to have two Xs for the same discipline/elements, which means two different ICD but different leaders. Depending on project complexity, this could happen.

Systems engineers will tell you that this approach is not enough, and they must develop further interface matrices. That is okay and perfectly normal. Don't resist and let them do it based on your general matrix. But for our infrastructure co-workers this is more than enough, in fact, probably too much, and they are not even used to this. But this will help a lot in focusing efforts on "open minds" to the requirements management approach.

Other Requirements: Environmental, BIM …

Until now, we have been covering the usual requirements for every Railway System, but these are yet again another opinion of mine.

As the Railway goes through multiple areas, it is normal to consider the environment. This does not only mean trees, animals, water courses, and contamination, but also community concerns. Grade-level crossings, noise increase on populated areas, childcare centers, schools, hospitals, retirement areas, or in general anything that the population considers an issue must be taken into account.

The problem is that the specialists in those fields are not used to the requirements approach and timing, so yet again, it is up to us to involve them early in the process. Nearly all environmental requirements will translate into something that your design team will have to consider. Noise walls, sedimentation ponds, climate-change studies, heat emission increases, and more. Explain to your environmental team that you want to have them included in your process and I assure you that you will make them happy. Now take advantage of that happiness and make them give you all requirements as SMART goals as early as possible. And of course, add those to your requirements database (immediately assigned to the relevant discipline as part of their design requirements).

BIM is another interesting point here. In the twenty-first century, it is normal to design in 3-D and have BIM (I will not go into explaining BIM in this book) applications for all the project departments, so the information is centralized and easily accessible though metadata. This also has a number of requirements mainly on how to organize elements, prepare them for data addition, document control coding, file sizes, geospatial coordinates, and more. Although this one is not actually related to the design or safety itself, it will help us with the information integration, which is key for the final safety case, as well as for operation and maintenance of the System, for accessing the huge amount of information that will be available in a seamless way.

There could be more requirements coming from other sources, but in all honesty, the ones I have described till now are the different sources of requirements I have dealt with in the past. I personally cannot think of any more, but I am sure the future will bring us more stakeholders with more requirements that can be treated exactly the same way we have covered to this point of the process.

Design Parties and Requirements. Procurement Recommendations

Do you think you can do all this with only one design company? The main designer that is with you at contract level?

Think again. I mean, you could, but you will end up paying twice. Here we have yet again another one of my personal views.

Each subsystem supplier is a specialist in what they do, and they will not share their patents and knowhow with anybody from the competition. As expected, they have their own products that they are interested in selling. This applies for every subsystem: access control, fire alarms, tunnel ventilation systems, train control, traction power, catenary, you name it. But this is not the case for infrastructure, right? I mean concrete, earthworks, drainage, reinforcement, conduits, and those things are pretty standard and not subject to trademarks and technological secrets except in a few cases.

I have seen in many projects, contractors not knowledgeable enough on subsystems, not recruiting the required people early, and then allowing the main designer to do it all till 100 percent design progress. Then, going on procurement process, only to discover that when questioning the market, a supplier says they will not provide what the designer is asking because their technology works in a different way, although complying with the original requirements.

This situation happens because the general designer can never be up to speed on what the market has evolved to and will use for their design what they have used in the past, which is always outdated or belongs to a specific company, type of technology, and approach.

An infrastructure technical director and project director going into a railway project for the first time will always make this mistake. We all think we know it all, so they will prepare their personnel forecast, adding subsystems people too late in the game. Even the testing and commissioning manager must be at the beginning, but we will talk about that later.

My recommendation: use your requirements team for this. Ask them to extract every single requirement you have pertaining to the subsystem you are interested in, and tell your designers to create a specification with those requirements plus anything else they might require. But, a performance specification. Be extremely clear and demanding on this. Only performance of the subsystem is allowed to be added on top of the requirements already included on your database. And this must be given to you no later than 60 percent into design development. You will soon understand why.

Now, get your performance specification plus your requirements and go to your procurement team to inquire in the market about a design-and-build contract. Add the condition that your subsystems supplier must be the detailed designer for that subsystem.

Each supplier interested must check the requirements list and answer if they will comply with them or not, and if not, why not. Also ask for an alternative for compliance or a proposed requirement modification that still allows for a safe solution at least (if not completely technically compliant, it must be always safe).

So, when the responses come back, you can select your design-builder based not only on price but also how far they are from complying with all technical and safety requirements. Worst case, you can even go back to your client and propose contract modifications that your selected supplier should have drafted for you. In that case, if your client says no right away, you know that the supplier cannot be selected.

Then, your main designer obligation will be to integrate that subsystem supplier design into the overall design. (Of course, you must have written all of this on your main designer contract in advance.) Each designer will still be responsible for its own design, but you will have an integrated product. Remember, you are the overarching Integrator (yes I know, capital I) of designs, so it is very important to have somebody on your team who understands and chases all these things, rather than the main designer, to ensure that everything works together. This is the role of the testing and commissioning manager, as we briefly described in the Roles and Responsibilities chapter.

To be able to do all of this, with you being the Integrator, you must have defined your interface's requirements, which is one of the key specifications that principal designers have to provide to you. Unless you clearly understand what it takes to integrate and make all those subsystems work together, you will fail because there is nothing worse than a "bunch" of subsystems working perfectly well on their own but unable to respond together.

One note: Some specialists like to have a dedicated role for this called integration manager, but I do not think it is necessary. In my view, the person who will do the testing is the one to do this role, as they will "suffer" everything in the real world at the very end of the project life and will have all the schedule pressure on them.

At the end, you will not have one design organization, you will have many. Each with their own contracts, conditions, peculiarities, way of doing things, legal approach, etc., and it is up to you to manage all of them to ensure an integrated, compliant, and safe solution.

And Now We Start design, Don't We? No! If You Do That, You Are Already Late.

Up to now, we have only talked about gathering requirements and organizing ourselves. Does this mean that we cannot start design until we have it all?

Well, in a perfect world, without schedule and cost constraints, that would be the case, but as you can imagine, if we do that, we have already failed. Everybody wants design to go on immediately, so it is up to us again to get as much information as possible to allow this to happen.

Early procurement, as described in the previous chapter, will allow us to relieve some pressure, but for this, we need an extraordinarily strong requirements-management team.

We should also aim to create during proposal time as many assumption documents as we can, to have some kind of baseline to commence with. During this time, we don't have access to clients, third parties, emergency services, or even train suppliers. (We are competing with other groups, so it is normal not to have that access.) There is nothing else we can do but to rely on our team strength, organize internal hazard identification workshops early, propose mitigations, and use past experience in engaging with third parties, emergency services, and operators. Of course, once you have the contract, contact all those groups ASAP to confirm (or not) your assumptions.

You must focus on getting as many requirements as possible to the main designers before reaching the end of the concept stage.

Another piece of advice: Make your designers produce basis of designs for each discipline and subsystem. Oblige them to think what they need, typical solutions, cross sections, and such. They will end up asking for plenty of clarifications, which are the requirements you will end up needing from all the parties we have been talking to until now. Establish your interfaces matrix, assign leaders, have the interface team establish relationships with third parties; ask questions, make assumptions, champion hazard-identification workshops, and push in general. It is only human nature to question everything once the mind starts to run. And we all need that initial push to have our minds really running.

Those first six months are going to be a living hell for the whole technical department, but it is the right approach. If not, too many redesigns, contractual fights later, and frustrations from everybody will occur.

Tough life. This is what you get for coming here. But it is a wonderful life that will expand your mind and improve your planning capabilities to levels you wouldn't imagine you could reach before.

Honestly, this is not for everybody. It is stressful and feels very lonely. Pushing so hard for so many things with lots of people not understanding why you are doing what you are doing so early in the game is a very ungrateful job.

Something Changes. What Do You Do? Configuration Management.

Now you have started your design. Ideally, you have multiple requirements from everybody, and you feel completely in control of the situation.

One day, somebody comes and throws a change at you. Different concrete strength, extra scope, change of maintenance approach, extra operation constraints.

Panic! What do we do? How do we inform everybody? What else this change will create? How do I estimate the cost and schedule impact?

In traditional infrastructure projects, in which we have all technical requirements and obligations crystal clear on our contract, this is the traditional change-control process with a change-control board. This process is normally led by the contractual department, who asks project controls to evaluate impact on cost and schedule.

In Railway systems, that is the perfect approach for ending in disaster. A change in one requirement could have a subsequent effect on other requirements and strategies. Remember the example of the maintenance road. If we were planning for it, and now we eliminate it, all our maintenance times will change (as we need to reach the place where the maintenance will occur with a rail vehicle so we lose working time); evacuations cannot use that parallel road, and emergency services cannot use it either. So, a large number of operational procedures, ways of responding to situations will have to adapt, and with those new approaches, our infrastructure will be modified to compensate for it.

This is what configuration management does. Starting from a requirement change (modification/addition/elimination), it analyzes the potential subsequent effects. With those in hand, it is needed to produce a gap analysis of what we might be missing and later, preliminary cost and schedule impacts, both in design and construction.

As you can see, it is far from simple, and lots of parties are involved in this. More complicated than the usual change control approach.

Somebody would think that with all those possibilities and studies needed, everybody needs to be involved in it, and to a degree, it is true. But we cannot simply stop everything we are doing just to analyze the potential impacts of a proposed (or already firmed and "must do") change. Once again, our team experience is key. A good configuration manager needs to work hand in hand with the requirements team to create, incredibly early in the game, based on experience, the kind of links between requirements that I gave with the maintenance road example. Which one is linked with which other requirement, procedure, strategy, so later, when the change occurs (and believe me, it will occur), it is easier to extract those relationships and consequences.

Once those relationships are completed, then it is time to ask a few questions of the specialists to supplement the initial assessment while in parallel engaging with project controls for a preliminary cost and schedule impact evaluation.

And now it is time to present all of that to our client. Yes, to our client.

Even if the scope increases, changes, or elimination is firmed and already instructed, we must present our analysis to the client, so the consequences of those decisions are understood and recorded. Keeping the approach that changes are only to be assessed by commercial teams is a huge mistake on Railway systems.

I have been able to revert firmed decisions by not-well-informed clients once they realized the extra cost and derived consequences of those decisions. On top of that, they always thanked me for that configuration analysis. We are here to help and support our clients, who do not need to have the same level of expertise and resources we do. If that was the case, we would not be needed.

This translates to happy clients who ask your opinion in advance when a change is coming because the level of trust will be higher with you than with their own teams. When you reach that stage, there will be no need to produce cost and schedule impact in advance to advise them, only the gap diagram and affection to other requirements and procedures.

This process, out of everything we are describing in this book, executed properly and systematically will be the one helping you the most with those difficult client relationships and will 100 percent increase the trust in the team.

How Can I Be Sure We Take Everything into Account? The V&V Process.

This is by far, my favorite part: the moment when everything comes together and makes sense. But also the most painful process there is. Thousands of hours, discussions, documents, corrections, last-minute rush, stress, adrenaline …

Lovely indeed!

But what is V&V? It stands for verification and validation. As usual, two words that will have different meaning depending on who you talk to. For infrastructure people, it means one thing, but for systems experts, it means something completely different. This is normally another case of conflict between those two worlds that it is up to us to throw a bit of common sense onto it.

Let's start with the reason for it.

As explained a few chapters ago, this whole assurance process was created thinking on integrating a high number of different subsystems together, to ensure that they all work together like an orchestra (I should have used this metaphor before).For not having many musicians playing together, what comes out of it will be good and pleasant. Most probably, it will be a disaster! We need a conductor, a director, to make sense of it. The same happens here; without an Integrator, the different subsystems will not work together, and the overarching System (in our case, a Railway) will not work, will not be safe, and we would have failed in our mission. All those chapters about requirements were trying to communicate to the different "musicians" (infrastructure and subsystems), what is needed from them, what impacts the others, to ensure that everything flows the way it should (again, a compliant and safe Railway System).

Now, let us take a subsystem, for example, a fire alarm (one of my favorites because both system and infrastructure people understand this one easily). Having a detailed design which ticks all our requirements (and I mean *all* of them) and proof from the installer that everything is in place is all we need, right?

Sadly, no. It also needs to work. So, here comes the testing, which we will discuss in an upcoming chapter.

Now that we have the testing, we are complete. (In reality, we are not, and I will explain later, but for this example, we are good enough. Please bear with me on this.)

On the just-explained example:

- Verification is our design and installation.
- Validation is our testing.

We apply the same with all our subsystems. Now, here comes the fight: systems-assurance experts' background is in subsystems, and they want to apply the same logic to everything on a Railway. This does not make any sense for infrastructure.

For example: We have a requirement that requests a crash wall for a speed higher than 100 km/h. By that logic, we should run a train until derailment and crash it against the wall. Completely bonkers!

So, for infrastructure elements (and this is my view on it), it should work this way:

- Verification is our design drawings and calculations.
- Validation is our ITP, quality audits, shop drawings.

That is the general logic, but reality is much more complicated. For the V&V process to be effective, both Vs must be independent from each other. You cannot have the same people doing both because the whole assurance process relies on control layers of independent verifiers and teams to ensure that nothing is left out.

Then comes the need for the V&V team and V&V manager. This team is the one in charge of checking that all documentation provided is responding to the specific requirement, but also the one in charge, at the beginning of the process, of identifying the artifact that will give definitive proof to the requirement. As we just explained, for subsystems, the definitive proof that everything is working is the testing, but for infrastructure …

Hence the vital importance of an experienced V&V manager. This person will lead the team that has to decide what is the final proof (formally called *artifact*) for each requirement, but also convince all the external parties involved in this process that these evidences are enough to ensure the compliance and safety of the Railway System. Experience again is key here.

To sum up the role of the V&V team: to establish the requirements artifacts at the beginning of the process and ensure the proof given is good enough and complete against each requirement. And in between, never-ending discussions with all parties about the artifacts, procedure creating, doc control alignment, and many more. You obviously will not need a full team from the beginning, but it is a fine balance to strike. A piece of advice: requirements leads are good for transitioning into V&V along the life of the project because they already know your requirements database inside out. A good V&V manager will naturally evolve into an even better assurance director, and those are exceedingly difficult to find and more difficult to keep.

If you noticed, I have started to explain the practicality of this process from the end; I have told you that you need a

V&V team who are the "police," the ones checking that verification is correct, complete, and enough for each requirement. And then, who and how do we do the verification? This is where the real challenge arises.

Two trains of thought (if you'll pardon the railroad expression): some assurance directors are of the view that it is better to have a complete independent team, with full access to doc control and that this team will link each document with each requirement, write a compliance statement, (compliance what?) and provide everything ready and even with a beautiful ribbon wrapped around it to our validation team. Others, like me, like to think about the vast amount of money that this independent team will cost, and we say that the generator of the proof of the requirement (artifact) is the one in charge of linking the documentation and the compliance statement in the requirements database.

In simple terms, this means that designers must be the ones linking their drawings to the requirements, construction to their ITP, and T&C teams to test results.

Easier said than done, because everybody is busy, teams never have enough resources, and overall, nobody cares about assurance because it is not that important—*so they say.*

This is again when a seamless link between document control, assurance teams, and especially our V&V team is key, as I will explain in the next chapter.

And now the cherry on top: the compliance statement. What is this?

In our verification process, we are pointing to drawings, calculation reports, test results, and all kind of documentation, but do we really expect that if our documentation proof is a two-hundred-page document, we simply say, "it is on document X" and that's it? For thousands of requirements? Well ... not really.

Believe or not, in every single project, designers, contractors, subcontractors, suppliers, quality control, and even doc control think that linking requirement to a huge document is proof enough and that the independent parties audit the whole thing while calmly and patiently read every single page of our reports to satisfy themselves (sarcastic mode off).

As in every audit, they might or they might not. And here is where the compliance statement comes into play. A compliance statement is nothing more than a summary or indication of where to look within our document to verify the requirement.

Let us use another real-life example: one of our requirements comes from a hazard identification, which requires that we must mitigate the possibility of a train derailing and hitting the columns of an existing road overpass. After doing statistical analysis based on other Railway systems in the world, our safety director and V&V manager have decided that the best mitigation is an engineering solution, infrastructure addition, rather than operational procedure. This means that it is up to our designer to come up with a mitigation for the hazard.

After analyzing the problem and consulting standards and other similar systems, the selected solution is to install jordan rail/derailment containment, rather than crash wall. (Jordan rail prevents derailment, protecting train, passengers, and structure, while crash wall is only protecting the structure.) Jordan rail is not effective in 100 percent of cases, but the statistical analysis shows that the remaining hazard is acceptable because although the consequence is catastrophic, the probability of it happening with jordan rail in place is very low.

We now have our jordan rail solution, and the standards even tell us the distance to commence installing it from the point we are interested in protecting and for how long. All of this is in our basis of design for track, which is around one hundred pages long.

As said, it is not enough under verification to simply link the requirement for structure protection with thc one-hundred-page document. The designer needs to write a compliance statement explaining why that solution is selected, and the standards used for it. Nothing long or complicated, but the designer takes responsibility for it (and here is where the North American liability approach comes into place, which we will discuss at the end of the book). Same logic for testing, ITP, or any other document we use. There needs to be a compliance statement that summarizes or indicates quickly how specifically the requirement is met.

If the requirement is repeated along the project multiple times because it affects multiple structures (in our example), then we must link it again with its own compliance statement.

For our projects, in which we repeat the same element many times, this can be tremendously tedious and prone to errors. Again, we rely on our team experience, which translates into our requirements team, V&V team, and doc-control team, which will be instrumental in this process.

I hope you now share with me the view that the V&V process is the most important one and as such, the one defining the whole approach. It is critical to engage our teams from the very beginning of the project, explain the needs to ensure that everybody—designers, contractors, suppliers, quality, T&C—understands what is required from them, how to make it happen, and what support will be given at all times. Let us not forget that in the technical department, we are a service to construction. We are the ones in charge of making their lives easier, not by doing their job but by establishing clear systems and procedures (and even tools if possible), that are easy to follow and do not oblige everybody to go back and redo things multiple times.

Even doing everything perfectly and ideally well, it will not be. There will be last-minute meetings, information missing, ITP/test reports unfinished, papers unsigned, temporary solutions to NCR or deficiencies not closed, and God knows what else. The level of stress is over the roof to close everything satisfactorily and on time because the number of days allocated on the overall schedule will never be enough. Even if it was originally, I bet you that construction or testing will be late (or both), with the consequence that the assurance process will have to be squeezed in and overlapped with something else, obliging the technical department to create something "imaginative" to make everything happen.

And I would not change a minute of it. The feeling of getting everything done at the end, crossing the finish line, and achieving substantial completion is great, and if you like these complex Railway systems as I do, you know what I am talking about.

Below is a summary table for doing V&V on infrastructure elements successfully.

	INDEPENDENCY OF TEAMS AS European Standards					
	V			&	V	
	Responsible	**Demonstration**	**Comments**		**Responsible**	**Comments**
Design	Design team	Calculations, Drawings, Feasibility reports	Doc control coordination and support		Validation team	V&V team performs second review and quality of evidence submitted to ensure compliance
Construction	Construction team	ITP, Concrete test, Quality inspection, audits to suppliers and subcontractors	EoR on site to ensure compliance with design as PEO regulations		Validation team	V&V team performs second review and quality of evidence submitted to ensure compliance
T&C	Construction team	Test reports. Test to be conducted by installer (self perform CJV, or subcontractor)	EoR on site to ensure compliance with design as PEO regulations EoR to review test report before submission to Assurance Team		Validation team	V&V team performs second review and quality of evidence submitted to ensure compliance

I will not be adding a similar table for subsystems because those processes are well defined already, and it is not the main objective of this book.

The Key Role of Document Control

I assume that by now, you have realized the huge amount of paperwork we are dealing with. (Digital paperwork; we are in 2023, when I am drafting this book and we avoid printing as much as possible), Creating requirements, minutes of meetings, ICD, concept of operations, ITP, design calculations, drawings, shop drawings, NCR, deficiencies, safety cases, ... NUTS!.

If we don't have a proper, efficient document-control team and system which everybody adheres to, it will not only be impossible to find everything during the verification process, but we will not be able to create interfaces between subsystems (remember our interfaces chapter?) or find all the documentation required for testing.

Go back to what we explained about WBS and SBS. WBS goes geographically, not by subsystems, and although it is great for costing, schedule, and construction tracking, it is horrible for testing and documentation finding when we are interested in, for example, all the structures, all the fire alarms, all lighting, and more.

A very efficient coding I have used during the years follows this structure:

Document No / File Number.:
Project Name - Originator - Location - Area - System - Activity - Type - Sequence

All codes must be three (3) digits except for Location (4) and sequence (5)

This coding allows the WBS approach (location and area) but also SBS (system, activity). It is particularly important that everybody adheres to it. This will become another teaching exercise for you and your team, which must include very clear procedures that everybody, even your client, must follow.

It means that not only is the coding key but also the workflows that go with it. ***Never use emails!***

There is nothing worse than transferring information with uncontrolled emails. People leave projects, come new, go on holidays, and information gets lost. Outlook, Word, Excel, and PowerPoint are amazing tools but have created a climate of complacency in which we just use those, as it is "easy" and we are used to them.

Those tools should only be use for work-in-progress (WIP) documents but never to transmit information to our subcontractors, consultants, or clients.

For official information transmittals, letters, drawings, reports, and ITP, anything and everything must go through document control with the appropriate workflow, which could have reviews in it or not, signatures, approvals, or whatever is contractually required (or even required under your own internal procedures). By doing that, document control becomes the custodian of the process, ensures that document numbers are assigned, used properly by us, contractors, suppliers, even our client. This way, documentation is stored by metadata and accessible to everybody in an organized manner.

The key here is to ensure that requirements are also categorized the same way as document number, and as you remember, our requirements database allows that and more. (Go back and reread the Requirements Definition chapter).

When we need to find all proof for a particular requirement—as the requirement will be also coded by location, area, system and activity—the verification process is easier for everybody, which in return reduces costs in personnel and lots of unnecessary frustration.

Now, add BIM into all of this, and tons of metadata, site reports, safety investigations, environmental and safety walks, and many more things that we want tied together.

Data management in our XXI century is the key to success but also the reason for failure.

Make doc control your best friend. It will be also difficult to set everything up, even more difficult for everybody to follow, but this is a do-or-die situation. Without a resolute document control team, processes and workflows, the project will fail. No middle point or shortcuts on this. The most difficult part: teach your client. Insist, insist, and insist. Many people on the client side will know and understand what you are talking about, and those are your allies and best advocates. Take them on board and ask them to help you implement everything that is needed to make your project a success.

Testing, Testing, Testing!

And now, we enter the unknown!

What is this guy talking about? I am sure, that if you come from the infrastructure world, your mindset is that when you install something that a supplier or subcontractor has been installing many times and it is certified, it works and that's it. Of course, you test it, but only a couple of times, and when everything does what it is supposed to do, job done and on to the next thing.

Again, wrong. Why? Because even if that is the way it should be (and you pay for it), that each individual subsystem does "its thing" unless it is installed incorrectly, is still not enough. Fire alarms, access controls, lights, HVAC, water pumps and similar subsystems, will work on their own, no worries. But what I am talking about is integration testing. Not the subsystems in isolation, but all of them working together so our "orchestra" plays the song in harmony. In this process is when things go funny.

No matter how many times you think somebody has done it, there is always a surprise. Because the combination of subsystems has never been used together before, because the software of one subsystem has an update and something that was working previously suddenly doesn't work anymore, because a particular situation that might happen occasionally nobody thought of, and when it does it fails. The possible reasons are endless.

But there is *one* reason, that always happens, and it is completely unacceptable: nobody bothered thinking about the integration since the beginning, and of course, nobody took the responsibility of doing it. Designers only worried about the different subsystems because they are different disciplines; procurement only gave subcontractors the scope for subsystem, not for integration; and management never thought in advance that this was required from the beginning, simply because if testing is so far away, why do I need a T&C team early? Let's save money on resources.

This is the biggest mistake, and it happens every single time. Again, different ways of thinking: the systems engineering side of this assurance world will say that an integration manager is a must, and they have to be independent from a T&C manager. This way of thinking is typical of systems engineering–based companies/consultancies, which specialize on subsystems for Railway systems, but they do not build, they subcontract. And they never do the infrastructure. I am talking about those who specialize in signaling, catenary, train supply, even in operations. I am not going to mention them, but you know which ones I am talking about. Their focus is design, systems engineering, and systems assurance. And they are excellent at what they do but have no idea of the whole cycle from beginning to end. They do not want to learn because it is not their business. Nothing wrong with that, but this creates clashes with our infrastructure way of thinking.

As this book is focused on the whole Railway System, including infrastructure as well as subsystems and from beginning to end, we are thinking on design, construction, testing, commissioning, and entering in operations.

Once more, I am going to give you my view on this topic: you do not need an integration manager. What you need is a testing and commissioning manager, and a good one. Somebody with site experience, who understands the millions of problems that happen in the real world, constraints with time, human behavior, working nights, weekends, fighting with designers to get signatures, with quality, with construction, even with the project director if needed. Someone who also knows what and how to negotiate on site for access, for testing with the construction teams, which will also be always late and don't want to help. That knowledge, that experience is invaluable. Because they can plan real durations for testing (which in every single project we underestimate), but also because they know what can go wrong in the integration testing itself from a design and practical point of view.

Because of all that, these people are like animals on the verge of extinction—very few and very precious. But also because very rarely they want to repeat many times that stress of testing a complex Railway System. What will likely happen is that you will find somebody with some site experience on a smaller Railway or part of it, and together you will learn the complexities of a bigger-scale one. And there is nothing wrong with that, because the instincts and integration knowledge of different subsystems are already there.

But—and this is the most important part—bring them at the very beginning of the design lifecycle, no matter what your budget, your company, or even your project director says. Put them in contact with the design team, review subsystems, ask questions, retouching procurement scopes to ensure that they flag everything needed later for testing, in terms of software, parts, certificates, diagrams, drawings, and specifications.

Be sure to provide them with everything they ask from you. (Of course, ask why, as cost is always important.) It is incredible the number of things that could and will happen when the testing phase commences, especially when integration testing starts.

What are the different types of tests? These can have many names depending on jurisdiction, country, and type of railway but generally speaking, the following definitions apply.

- Factory Acceptance Test (FAT): Very obvious. The subsystem is tested in its factory, in a safe and controlled environment before being shipped on site. I will not go into the paperwork detail, but of course, this will be required later for our verification and validation process.
- Site Acceptance Test (SAT): Repetition of the FAT but on site, after being installed. Remember, the FAT was performed in a controlled and ideal environment, while our site is not. It is dirty, cold, hot, there is construction at the same time, and things are halfway done.
- Global Site Acceptance Test (GSAT) or Integration Testing: The funniest part. Making multiple subsystems work together and ensuring that it all works. Here is when things can go south very quickly and dramatically. Typical examples of this are the interaction pantograph with catenary, train wheel with track, blue light system (BLS) for cutting power, fire alarms opening fare

gates while blocking or unblocking elevators (this goes for a full book, but I will talk a bit in the next chapter), and many others.

For each one of these tests, paperwork must be spotless, signed properly without caveats or disclaimers because this is where we really bet the success or failure of everything. No matter how nice and well executed our bridges and tunnels are, at the end of the day, if the trains do not run or in an emergency, things are not doing what they are supposed to do, our System will not open, and it will be a failure. Remember what I explained at the beginning about the difference between a road bridge and a Railway System, because that is the key in all of this.

With this, we are nearly done. At least on site. Everything is built, subsystems tested with positive results, and it is integrated and ready to roll.

But we are still missing the paperwork, and as an incredibly wise manager told me many years ago, "Your project is only finished when the amount of paper is bigger than the amount of concrete." I did not understand it at that time, but now I know how right this gentleman was.

Even when you think you have succeeded because everything works, you are not done without all certifications, signatures, and approvals required. That is precisely the main problem in North America, and one of the reasons why every single Railway System is late or over budget or both.

Next chapter could be extremely boring, but it is needed. I will try to avoid being tedious but if it is, at least you can read it before going to sleep and be grateful for finding a cure for your insomnia.

North America vs. the World. Legal Framework Differences. An Option to Deal with It.

This chapter is a special one for me. Not because I love the topic, which I certainly do not (by far), but because it is the one that creates more anxiety, misunderstandings, discomfort, delays, and very unnecessary and nasty fights between good people, which is never needed. All because of a lack of understanding and communication, which is often the core of all issues in our business.

The assurance process was originally created in Europe, as I said at the beginning of this book, in an effort to ensure that nothing was left out on projects, mainly on manufacturing ones, including multiple subsystems that need to perfectly "talk" with each other for the whole element (System) to work. As usual, the best example is an airplane. Later, this process was extended to us in the Railway world.

It is based on a series of gates and checkpoints, which are nothing more than reviews by groups of expert people in the relevant field. Although I am simplifying a lot, the core of it is based on that. Taking a simplistic approach, we could reach a level of design, stop everything, give design to an external group of experts who will review it and tell us if we have covered everything possible in terms of compliance, interfaces, constructability, safety, security, cyber, human factors, and so on. If everything is good, this group will give approval, and we continue. Repeat this process a couple more times during the whole design progress, and then, when everything is finished, we can build. Similar logic applies to construction, later on testing, commissioning, until the final review of all documentation (this is the safety case that we will be touching on later) is completed. Once we have the final approval based on all processes previously described, we cannot start using the Railway.

As this will take an extremely long time, we created all the processes and procedures described until now to overlap activities, track and correct, so we can design, construct, and test different project stages and areas at the same time. (We are building in one place, testing in another, and even only at 50 percent of the design progress somewhere else.)

You have probably noticed that everything is by "group of SME (subject matter experts)," not by an individual person. And here lies the root of the problem.

Anywhere in the world but Canada and the United States, responsibilities are by companies, not by individuals, unless it is demonstrated that something happened because a particular design (or construction or test) was done incorrectly or even not done, on purpose to save time, money, or something else. A negligence was committed. But if that same situation happened because of a mistake, the responsible party is the company, not the person, and it will be under civil law, not criminal law. This means that the penalty will be monetary, loss of practice for an amount of time in extreme cases, but not prison for a particular individual. That is why all companies have many layers of quality control, in which everything is checked by another person or group before being considered finished. This applies to designs but also to creating or updating new standards that design companies must follow. Those standards are approved by a recognized organization and not signed by anybody but that organization. Same with designs: they are signed by the companies, not by the individuals. What this does is make the company responsible for not only having a compliant design but also demonstrating that these design companies have followed their own quality controls, which all need to comply with ISO 9001, meaning that they are auditable.

When you apply the assurance approach of having all requirements given to those designers early, the companies become responsible for applying them, and the same quality control procedures are applied to those.

The same level of reviews, responsibilities, and even homogeneity, which also creates a level of comfort on how the work will be done.

In Canada and the United States, the responsibility for a mistake lies with the individual, not with the company, and it has criminal consequences. Designers sign their drawings, standards are signed by people (even when it is a group), permits, work procedures, everything.

So, if the same mistake we were just talking about on a design happens in these countries, the designer who signed the drawing will be responsible for it, and if it ends in an accident, will be faced with criminal charges. That designer alone, not the company the person works for. This, in consequence, translates into design companies not always having robust quality control procedures and with as many layers as everywhere else, because why do they need it? The responsible party is the designer, so this person better be sure of the work they do.

What would you do if you were left alone (or nearly alone) with that responsibility and if by making an honest mistake, there would be the possibility of losing your license, which is what you need to work and feed your family? And on top of all of that, even the possibility of losing everything you have and ending up in jail?

You will ensure that it will not happen by checking a million times, asking plenty of questions, including lots of exclusions, only including in your drawings what you have designed, nothing else, and even overdesigning in some cases. That is why here, drawings are only showing portions of the whole design, which makes it difficult to purchase and build, but worst of all, to integrate the different elements coming from infrastructure and subsystem designs.

That is why we try to go to 3-D element design, so everything is included on the overall Federated 3-D model, but this approach creates other problems.

- 3-D models are not recognized for signature in North America.
- As very few government agencies, cities, or towns are used to 3-D design, there is a serious lack of expertise on 3-D modeling and BIM.
- In case of any discrepancy between the 3-D model and drawings, only the drawing is valid (legally recognized), so we run the risk of having a wrong integration.

Now, having this critical legal difference in mind, imagine our assurance process. Designer focus is on compliance, not safety derived from hazards or interfaces, not because they do not care but because their first obligation is the legal element. As long as that designer complies with standards and contracts, they are safe. A missing interface between subsystems becomes a challenge if identified late and the process of responsibilities, independent drawings, far too many questions, and so on, starts again.

Remember the compliance statement on the V&V process that we said should be a clear indication or summary of how the requirement is demonstrated? It's another possibility for those designers to make a mistake, so they simply refuse to do it. Risk management in its most basic form: if you do not do it, there is no risk.

That is why every single request, every change, every comment becomes a fight, frustration, constructor not understanding and systems engineering people, mainly coming from other countries not even believing the reluctancy to all of it.

Just try to be in the shoes of whoever is signing something and think again what you will do.

This is simply why every single Railway System in North America that has failed so far in the last few years are the ones in which assurance has really become so important: responsibility, different legal framework, and trying to impose a European approach into North America without analyzing the situation here.

For North America, it is required that the design be signed by a professional engineer recognized in the territory in which the Railway is being built (or upgraded, if existing). But that signature is only for compliance, not for safety or interfaces. For example: a professional engineer will sign that in its design, a room has the required number of water sprinklers under the building code but will not sign that it is safe in case of fire because all identified hazards have been mitigated. Simply because those hazards may not be included in the building code. Again, the best way to avoid risk is not to take it in the first place.

Then how do we fill the gap? One way would be to convince all designers to do the same as Europe, but let's be honest—unless the legal framework changes completely, it will not happen. As much as we complain that it makes no sense, the legal obligations are what they are, and it is better to adapt than trying to change them completely.

The approach I have taken in my career is simply to take that responsibility on myself. As the person responsible for the technical department, I had the designers, temporary works, testing, safety and assurance with me. This means that I am the Integrator of all of those (another word with capital letter) elements, as well as ultimately responsible for all requirements and interfaces to be incorporated in our design. This does not mean that you relieve designers of their legal responsibility but only that you cover the gap they have not. In the previous sprinkler example, the safety team would have done a robust hazard analysis, identified all hazards possible for the particular situation, and suggested mitigations to cover them. Those mitigations translated into requirements, which landed into our requirements database, were categorized, and subsequently, I would have instructed the designers (which means variation to original contract, with all that this implies, as it is a scope increase) to take all of those into account. Then the V&V team checks that all requirements are demonstrated through calculations, drawings (I know I am missing the compliance statement, but we will go into that soon) and later built and tested. Finally, with all that documentation in my hands (technical compliance signed by the designer and demonstration for all other requirements through the process), I would join all the pieces together with a letter indicating that it is compliant and safe (with the wording obviously more cautious than that) based on our assurance process.

Still, it is a big risk for the technical department leader, but it goes with the title.

Now, extrapolate the example of the sprinklers to the whole Railway System for all requirements coming from all parties: interfaces, environmental, hazards. It is very tough, demanding, and a huge amount of stress. That is precisely why you have seen me insisting so much in doing so many things at the very beginning and having competent people also very early in the project, not only a few months before we think they are going to be needed for construction.

Ask yourself.

- Would you take that one?
- Would you do the same?

If answer is no—and there is nothing wrong with that—my recommendation is to take another job than the technical department lead in a Railway System. Or if you can find another solution to breach that gap in North America, please let me know. Nothing will make me happier!

Manuals and Procedures

We are talking about Railway Systems with lots of subsystems, bridges, tunnels, drainage, and a multitude of different elements such as computers, servers, switchgears, transformers, lights, and many more.

Every single one of those has different durability, pieces to be replaced, maintenance procedures, even cleaning needs.

And of course, the same type of product will have diverse ways of being put together in the factory, pieces, elements, depending on the manufacturer. But the same logic applies with infrastructure elements such as rails, sleepers, ballasts, even tunnels and bridges.

No matter what our contract says about durability requirements (one hundred years, twenty-five years, or whatever), every single infrastructure, subsystem, and component needs to be maintained, and as we explained in our RAM chapter, we must consider how to do that maintenance and how much time is needed to do it.

For this, we have the manuals. And those are exactly what you imagine they are, similar to the one for our car: description of components, how to clean, maintain, how often to replace each component, and why.

Ideally those manuals have to be produced by each supplier we have engaged or our designers (for bridges, track, tunnels, and general infrastructure) and also ideally, they should have a consistent "flavor" and writing, taking into account our RAM analysis, access roads, specific infrastructure built to facilitate tasks, and even a link with our FRACAS procedure (In two chapters we talk about it.)

But the reality is that if we simply add in our procurement package that suppliers must produce manuals for their products, they will do it, but without considering anything of what I have just described. So, it falls once again on us. There is nobody better for this task than our testing and commissioning group, because they are the ones who have been providing input to design, helping the procurement development, and adding requirements to facilitate their own obligations. Yet another reason for having these people early in the project.

The idea can never be to oblige our suppliers or our designers to know everything as well as we do, thinking that they can simply write the manuals for us. The focus should be on creating the provisions that will allow us to include or adapt the manuals they will provide into an overall manual for each component. This overall "manual of manuals" will be the one considering everything that is needed and creates that consistent “flavor” we were talking about a few paragraphs before.

One last comment: one manual per structure or component. Never do one big manual with chapters. Although that approach could be good years ago, in these days with BIM and digital twins available, it is better to do multiple manuals. Why? Because in that way, it will be very easy to link each manual with its component on our BIM model, so when the maintainer clicks on the element, all documentation related to that element, including the manual, will be easily available for them.

Of course, it will be up to our combined experience as to which level we want to break the manuals into. A bit of common sense will have to prevail, but I always recommend having this discussion very early in the process with the maintainer and the client in the same room. Yet again, record and submit the agreements to all parties. You will avoid many unnecessary complaints later.

Wrapping It Up. The Safety Case

"There are many ways to skin a cat", and equally, many ways of putting a safety case together.

This is another difficult-to-understand term, but in reality, it's very simple. So, what is a safety case? It is the document that demonstrates that our Railway System is safe to open for operation.

"Thanks for stating the obvious and not explaining anything at all!"

This is the answer you will be given by nearly everybody in the systems assurance world, because for them, it's pretty obvious, but how do we know that we are doing it properly if there is not a clear template on how to do it? No idea, in all honesty.

But I found a way, which I have been describing to you in this book without you noticing about it. And no independent party, client, or government body will be able to reject it. But only if all the information is included in a clear and concise way (pretty obvious but based on the journey we had till now, not that easy).

The answer is requirements management and my personal obsession of making everything a requirement—a SMART one, easy to quantify, demonstrate with documentation, and explain. Think about it: if any safety need is transformed into specific requirements, as we have learned so far, and you have tracked down everything and attached all relevant paperwork, nobody can tell you that you are unsafe.

Each hazard has its mitigation that has been transformed into specific requirements. And you have your drawings, construction paperwork, tests, and integration tests linked to them. You have everything in place for an audit, you are able to answer questions, add more detail if required, but most importantly, put together your safety case relatively quickly.

This is my template: Your safety case will be the typical first pages of introduction, what and where is your Railway System, its objectives to improve communication and traveling in the area where it's built, your client, benefits of all kinds. Then you will have your safety team explaining about your hazards-identification sessions, security analysis (TVA, if you remember), cybersecurity, stakeholders involved, and most importantly, how you decided to go with the SMART requirements strategy for all of them, be it technical, interfaces, environmental, third parties, hazards, all of it. And how for your team, no requirement is more important than any other, and that is why they are all treated the same way in your requirements database but clearly classified, so you can quickly pull out the different categories, including all safety-related ones.

Then continue explaining your configuration-management approach to track changes as design of infrastructure and subsystems evolve, and finish with your verification and validation process (V&V) as detailed in our V&V chapter.

This will be your safety case body.

Then you create appendices with hazard logs and TVAs, their requirements, and show that all of them are closed, demonstrating that all hazards are mitigated.

Finally, as proof of all the above, you include the whole V&V requirements and proof list, classified by categories (safety, security, cyber, interfaces, technical, etc.) with links to your requirements database and in that one, links to all documentation in your document-control system.

Now you understand why this chapter is called "Wrapping It Up."

By having rigorous requirements, V&V, doc control, an experienced team from the beginning to identify requirements, create processes and coach the rest of the construction teams you will succeed, and your final job will not be that painful.

Don't get me wrong. It will still be incredibly painful because things are never ideal, documents are not signed, there are testing delays, last-minute additions, even IT problems and many more you will discover on your own.

But at least you will be ready, with a clear strategy, systems in place, and people willing to row together in the same direction, understanding the final goal.

Far too many projects only go into this in their last year, and it is impossible to have it done on time.

I know, I know, easier said than done, but it is your job to convince whoever you need to have everything ready from the beginning my friend.

Continuation. The FRACAS Process

What more is there to say if we have successfully submitted our safety case, gotten it approved, and even handed over the project?

No testing in the world can prepare you for what comes next: people.

What?

Yes, people. Operators, maintainers, passengers, and even politicians.

No matter how many tests, manuals, rules, or even automatic subsystems you create, there will always be situations for which you have not predicted and will surprise you.

An operator moving a train back into its storage area without noticing that somebody left a piece of plant too close to the track that scratches the whole train side, leaving it unusable.

A passenger who arrives at the train at the last minute, and instead of waiting for the next one to come, decides to jam the doors open and stops the whole System because train has detected a door failure.

A politician promising to increase the number of trains to reach System maximum capacity only a few months after opening.

The examples are endless with our infrastructure and subsystems in warranty for a few years, and we must respond to all those unexpected events. (Well, maybe not the political promise, but who knows.)

This is where failure reporting, analysis, and corrective action system (FRACAS) comes into place. This mouthful name is nothing more than a way of recording failures and trying to discover how to fix or bypass them. (Yes, the solution is not always to fix.)

First, we need to understand if the error is a "one-off" or repetitive. If something happens, and by fixing it, it does not happen again, problem solved. It gets recorded, and we move on.

But if after fixing that failure, it happens again and again and again, then we have a systematic problem that simply was not thought about in advance. This is when a fix might mean a complete redesign and reconstruction, which is not always a good solution because our System is already operating. In this case, we will need to engineer a solution, which could be infrastructure additions, software updates, operational procedures, or even in extreme cases, shutting down—partially or completely—

the System because that repetitive error could end up in a fatal situation. (This is an extreme case and exceedingly rare.)

That is the FRACAS, a reporting system to record, categorize everything unexpected, and identify the repetitive problems that require an action to be taken.

It is our obligation as System design-builder to have a team during the contractually agreed period (normally the warranty period) to perform FRACAS analysis and decide if the responsibility to correct is on us, subcontractors, operator, maintainer, or even the client. And of course, to ensure that it is satisfactorily implemented in our System with the same process we have been describing in this book.

Obviously, this is far from ideal, so it emphasizes again the need for experienced team members from the very beginning, to consider many of these initially assumed unknowns (as they have suffered them already) and decide with all parties if we must consider them as part of our design (so more requirements) or not.

The End?

In theory, my friends, this is the end, but it never is, is it?

In the real world, we will have our client calling us with complaints, questions, additions, further information requirements, changes, and in some cases, we will have to deal with claims, mediation, and all those things that should not happen but sadly always do. And if not with our client, with suppliers, designers, or contractors. Many people participate in bringing a Railway System to life, and each one of them has their own interests, personalities, agendas, hidden agendas, and needs.

We have to deal with all of that, but let's be honest: you wouldn't be reading this book if you already didn't know all of that and at the bottom of your heart, you really didn't like the whole experience, with its ups and downs, joy and gloom moments, stress, failure, and finally success.

And you will be doing it all again, with more years, more experience, more lessons learned, more contacts and people you would love to work again, and others you would not.

This railway world is surprisingly small, and it never ceases to amaze me that I either found lots of people over the years that I would have never thought I'd see again, or a new colleague arrives who knows somebody or worked with somebody who was with me in an old project, and we end up connecting again, even remotely or by email. Then the relationship starts all over again, sharing moments, memories, new gained expertise, making everything worth it and the reason for not leaving this frantic world for something easier, such as putting out fires with your bare hands or piloting a plane without working engines.

I hope you all enjoyed this small guide, my first book, as much as I have enjoyed writing it and remember what I said at the beginning: if you want to join our small crazy community, find me on LinkedIn, through a message and let's start "moving the pieces".

Come to the dark side of the System!

About the author

Ignacio (name he prefers to go by) is a Civil - Structural Engineer and experienced Project Director with over 20 years of experience in large-scale EDPM, P3 and Design-Build infrastructure projects in Europe, Australia and Canada His career is particularly focused in Railway Systems, both freight and mass transit, including high speed Railway Systems.

Ignacio started his career overseeing complex temporary works design departments in high-speed rail projects in Spain to move later onto construction, focusing on tunnelling and elevated structures. With his family, he moved to Australia in 2010 and became Construction Director for Sydney Metro elevated portion and signature cable stay bridge and in 2016 joined the Australian Rail Track Corporation as a Project Director for the P3 portion of the Inland Rail program.

Since his arrival to Canada in 2018, he became Project Director with SNC Lavalin and later in 2019, Ignacio moved to Aecon and joined NouvLR as Technical Director on the REM project in Montreal till August 2022. He then has joined ONxpress as Technical Director for the GO Expansion Program, being this one his current role.

Appendix. The graphs

The following pages are simply some of the graphs used as reference on this book on bigger format (one graph per page) so if needed, they can be used or consulted in your project.

I hope you find them useful.

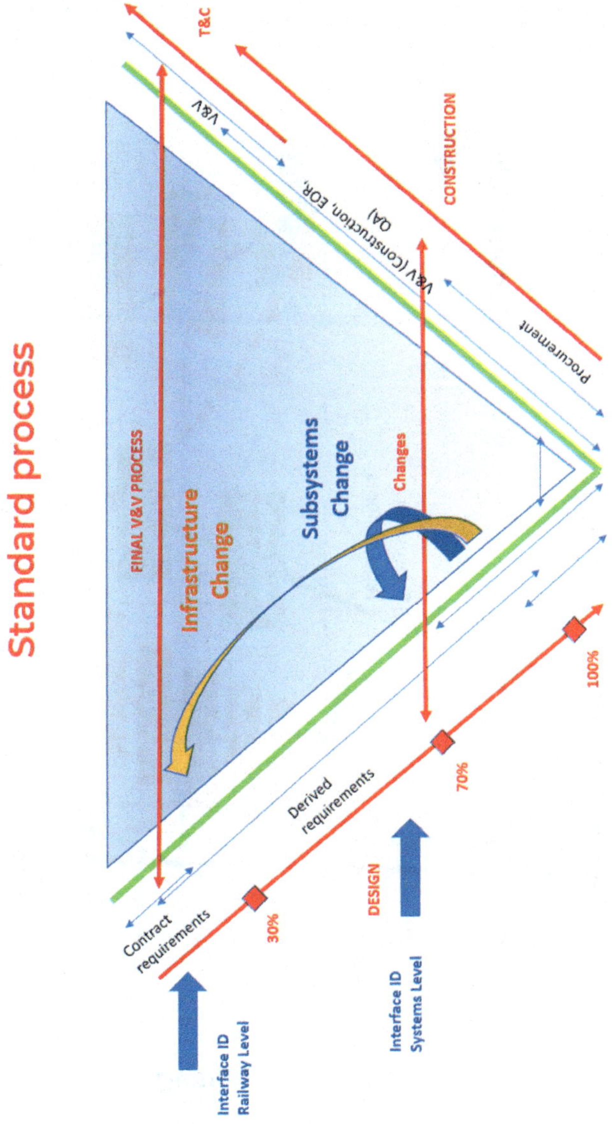
Standard process
T&C
V&V
V&V (Construction, EOR, QA)
CONSTRUCTION
Procurement
FINAL V&V PROCESS
Infrastructure Change
Subsystems Change
Changes
Contract requirements
Derived requirements
30%
70%
100%
DESIGN
Interface ID Railway Level
Interface ID Systems Level

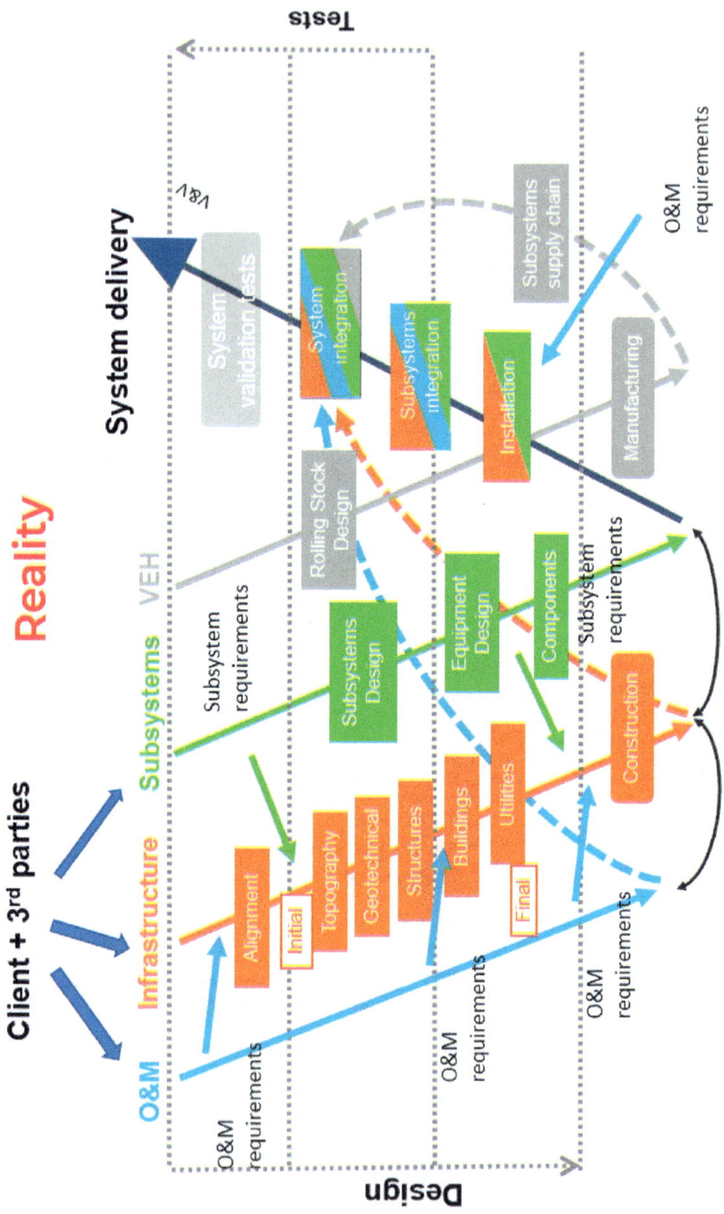
Client + 3rd parties
Reality
System delivery
O&M
Infrastructure
Subsystems
VEH
Design
Tests
V&V
O&M requirements
Alignment
Initial
Topography
Geotechnical
Structures
Buildings
Utilities
Final
Construction
O&M requirements
O&M requirements
Subsystem requirements
Subsystems Design
Equipment Design
Components
Subsystem requirements
Rolling Stock Design
Manufacturing
Installation
Subsystems integration
System integration
System validation tests
Subsystems supply chain
O&M requirements

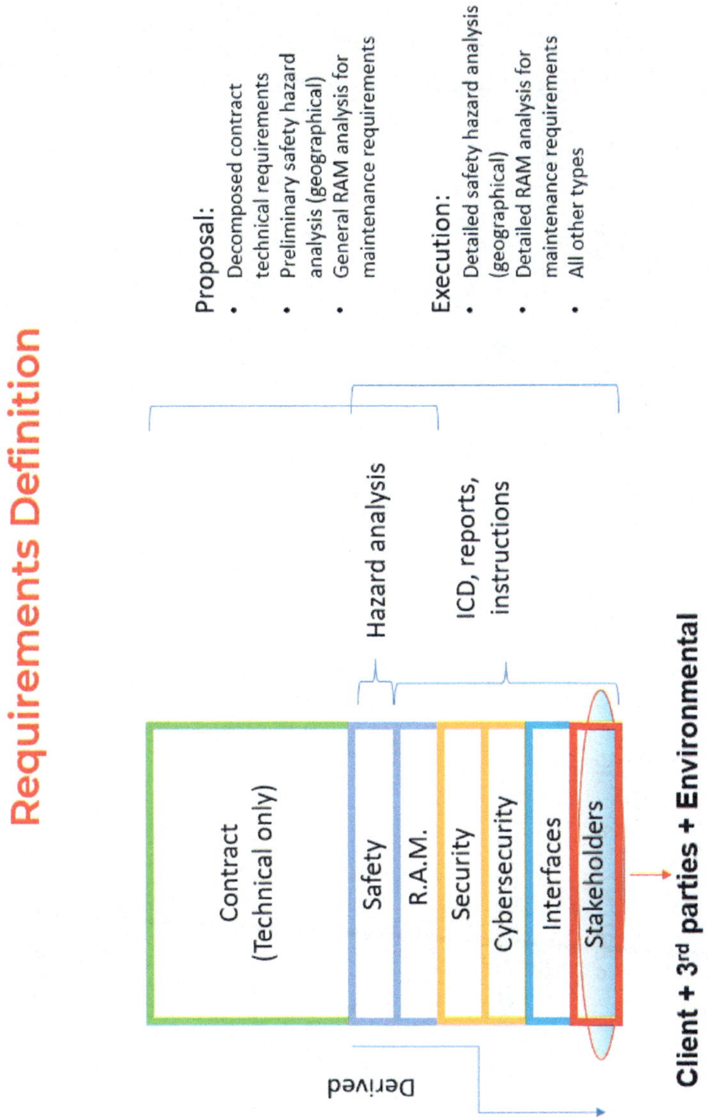
Requirements Definition
Contract (Technical only)
Safety
R.A.M.
Security
Cybersecurity
Interfaces
Stakeholders
Derived
Hazard analysis
ICD, reports, instructions
Client + 3rd parties + Environmental
Proposal:
Decomposed contract technical requirements
Preliminary safety hazard analysis (geographical)
General RAM analysis for maintenance requirements
Execution:
Detailed safety hazard analysis (geographical)
Detailed RAM analysis for maintenance requirements
All other types

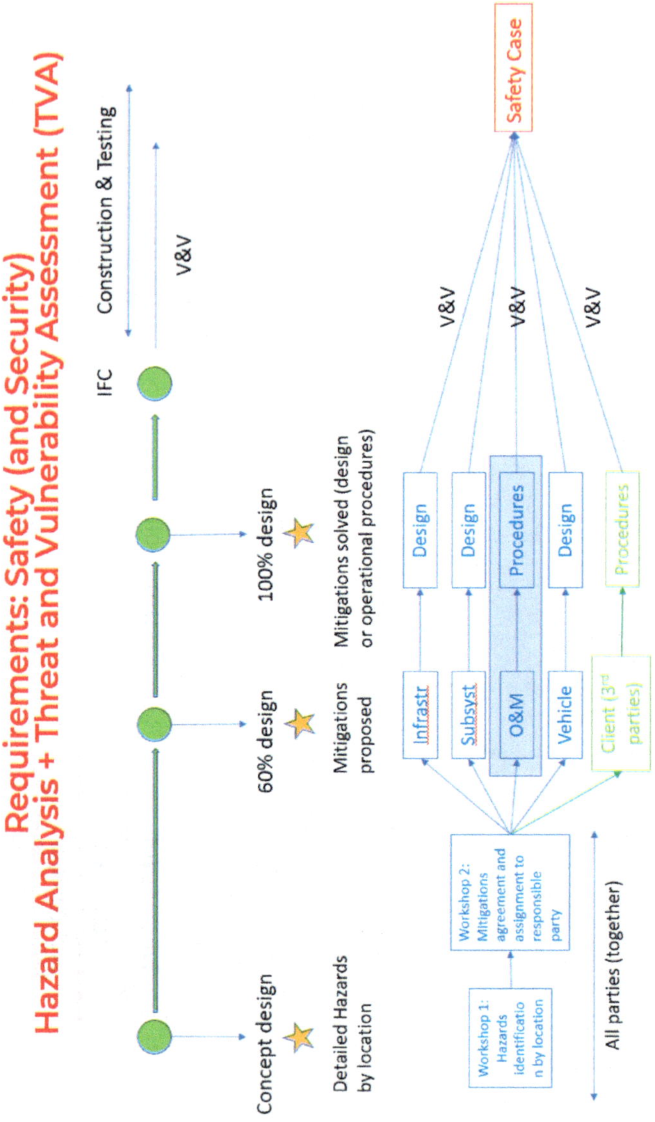
Requirements: Safety (and Security)
Hazard Analysis + Threat and Vulnerability Assessment (TVA)
Concept design
60% design
100% design
IFC
Construction & Testing
V&V
Detailed Hazards by location
Mitigations proposed
Mitigations solved (design or operational procedures)
Workshop 1: Hazards identification by location
Workshop 2: Mitigations agreement and assignment to responsible party
Infrastr.
Subsyst.
O&M
Vehicle
Client (3rd parties)
Design
Design
Procedures
Design
Procedures
V&V
V&V
V&V
Safety Case
All parties (together)

Group	No.	Element
Evacuation	1	Evacuation
	2	BLS
	3	Access/egrees
	4	Access control
	5	Fencing
	6	Illumination
Interstation	7	Alignment
	8	OCS
	9	Track
	10	Stray current
	11	Cable routing
	12	Drainage
	13	Earthworks
	14	Utilities
	15	Structures (new/existing)
	16	Geotech
	17	Noise attenuation
	18	Scada (lines as needed)
	19	Fiber Optic
	20	Medium Voltage
	21	Lov voltage
	22	Ventilation
	23	CCTV
	24	Intrusion
	25	Heavy Object Detection
Stations/MSF/Substations	26	Architectural
	27	Struct (buildings)
	28	Mechanical (buildings)
	29	Electrical (buildings)
	30	Landscape
	31	Passeng Announcement
	32	PSD
	33	Passanger communication
	34	Fare Collection System
RSSOM	35	CBTC
	36	Signalling
	37	Train communications
	38	Operations & Maintenance
	39	MSF equipment
	40	Rolling Stock
	41	Client
External/3rd parties	42	Ministry of Transportation
	43	Building Authority
	44	Municipalities
	45	Firefighters
	46	Police
	47	Ambulance
	48	Ministry of Environment
	49	Others (CN, CP, EXO,...)

	INDEPENDENCY OF TEAMS AS European Standards					
	V			&	V	
	Responsible	Demonstration	Comments		Responsible	Comments
Design	Design team	Calculations, Drawings, Feasibility reports	Doc control coordination and support		Validation team	V&V team performs second review and quality of evidence submitted to ensure compliance
Construction	Construction team	ITP, Concrete test, Quality inspection, audits to suppliers and subcontractors	EoR on site to ensure compliance with design as PEO regulations		Validation team	V&V team performs second review and quality of evidence submitted to ensure compliance
T&C	Construction team	Test reports. Test to be conducted by installer (self perform CJV, or subcontractor)	EoR on site to ensure compliance with design as PEO regulations EoR to review test report before submission to Assurance Team		Validation team	V&V team performs second review and quality of evidence submitted to ensure compliance

Made in the USA
Coppell, TX
05 November 2026

11302977R00075